AF769131

IntechOpen

Supporting open minds since 2005

Recent Advances in Nanophotonics - Fundamentals and Applications
http://dx.doi.org/10.5772/intechopen.87904
Edited by Mojtaba Kahrizi and Parsoua A. Sohi

Contributors
Samir Kumar, Prabhat Kumar, Anamika Das, Chandra Shakher Pathak, Arash Ahmadi, Andreas Fognini, Michael E. Reimer, Anand Bhatia, Sukeun Eom, Min-woo Kong, Kwang-seok Seo, Thamer Tabbakh, Khursheed Ahmad, Mohd Quasim Khan, Mojtaba Kahrizi, Parsoua A. Sohi, Vishal Sorathiya, Ankur Pandya, Sunil Lavadiya

© The Editor(s) and the Author(s) 2020
The rights of the editor(s) and the author(s) have been asserted in accordance with the Copyright, Designs and Patents Act 1988. All rights to the book as a whole are reserved by INTECHOPEN LIMITED. The book as a whole (compilation) cannot be reproduced, distributed or used for commercial or non-commercial purposes without INTECHOPEN LIMITED's written permission. Enquiries concerning the use of the book should be directed to INTECHOPEN LIMITED rights and permissions department (permissions@intechopen.com).
Violations are liable to prosecution under the governing Copyright Law.

Individual chapters of this publication are distributed under the terms of the Creative Commons Attribution 3.0 Unported License which permits commercial use, distribution and reproduction of the individual chapters, provided the original author(s) and source publication are appropriately acknowledged. If so indicated, certain images may not be included under the Creative Commons license. In such cases users will need to obtain permission from the license holder to reproduce the material. More details and guidelines concerning content reuse and adaptation can be found at http://www.intechopen.com/copyright-policy.html.

Notice
Statements and opinions expressed in the chapters are these of the individual contributors and not necessarily those of the editors or publisher. No responsibility is accepted for the accuracy of information contained in the published chapters. The publisher assumes no responsibility for any damage or injury to persons or property arising out of the use of any materials, instructions, methods or ideas contained in the book.

First published in London, United Kingdom, 2020 by IntechOpen
IntechOpen is the global imprint of INTECHOPEN LIMITED, registered in England and Wales, registration number: 11086078, 5 Princes Gate Court, London, SW7 2QJ, United Kingdom
Printed in Croatia

British Library Cataloguing-in-Publication Data
A catalogue record for this book is available from the British Library

Additional hard and PDF copies can be obtained from orders@intechopen.com

Recent Advances in Nanophotonics - Fundamentals and Applications
Edited by Mojtaba Kahrizi and Parsoua A. Sohi
p. cm.
Print ISBN 978-1-83962-843-6
Online ISBN 978-1-83962-844-3
eBook (PDF) ISBN 978-1-83962-845-0

For EU product safety concerns:
IN TECH d.o.o., Prolaz Marije Krucifikse Kozulić 3, 51000 Rijeka, Croatia, info@intechopen.com or visit our website at intechopen.com.

We are IntechOpen, the world's leading publisher of Open Access books

Built by scientists, for scientists

5,100+
Open access books available

126,000+
International authors and editors

145M+
Downloads

156
Countries delivered to

Our authors are among the
Top 1%
most cited scientists

12.2%
Contributors from top 500 universities

WEB OF SCIENCE™

Selection of our books indexed in the Book Citation Index
in Web of Science™ Core Collection (BKCI)

Interested in publishing with us?
Contact book.department@intechopen.com

Numbers displayed above are based on latest data collected.
For more information visit www.intechopen.com

Meet the editors

Mojtaba Kahrizi is currently a Professor with the Department of Electrical and Computer Engineering, Concordia University, Montreal, QC, Canada. He has authored or co-authored over 200 journal and conference articles. His research spans a wide spectrum under the umbrella of material sciences, solid-state devices, microelectromechanical systems, and nanoscience and nanotechnology. He has been actively involved in investigating the fundamental issues related to micro and nano structures and designing and fabricating micro/nano devices with applications in biomedical, sensors, and energy storage. Most recently, he is involved in research to develop low-cost fabrication of semiconductor nanowires, such as silicon, ZnO, and TiO2 with applications to develop greenhouse gas detectors, hydrogen storage, and detection of micro molecules in water.

Parsoua Abedini Sohi received her M.A.Sc. and Ph.D. degrees working in the field of nanoscience and nanotechnology at the ECE Department of Concordia University, Montreal, Canada. She is currently pursuing her research activities as a postdoctoral fellow at the Centre de Collaboration MiQro Innovation (C2MI). The fabrication and characterization of novel nanostructured materials, the development of methodologies and process flows for creating these structures and applying the fabricated structures in miniaturized sensing devices are part of her research interests.

Contents

Preface

The emerging nanophotonics from optics and nanotechnology deals with interaction of light with matter at the nanometer scale. It is the subject of much research and it covers many disciplines in science and engineering. As it is impossible to focus light at the nanometer scale using conventional optical components, we need other techniques like surface plasmonic, metal optic, and metamaterial to confine light at the sub-wavelength scale. The discovery of new phenomena based on the quantum principle, development of a large number of nanostructures and nano devices, and the vast application of nanophotonics have created numerous opportunities for new research areas both on scientific fundamentals and technological applications. Nanophotonic structures and devices are the promising future technology that can revolutionize, to name a few, the optical communications, energy conversion, nanoscale optical sources, integrated information systems, image processing, and sensing devices in all disciplines particularly in nano-medicine. Studies and research in the area of nanophotonics include the fields of, but not limited, plasmonics phenomena and its application, nano sensing devices, nano waveguides, nano optical sources such as LED and LASER, photonic crystals, and quantum optical devices. This book aims to present a collection of research articles on scientific fundamentals, fabrication, and characterization of devices based on nanophotonics, and on related fields.

The book presents 8 chapters divided into 3 sections. The focus of Section 1 is on quantum devices and it contains two chapters. Chapter 1 describes the fundamentals of nanoplasmonics with details of SPR and LSPR followed by their applications. Chapter 2 discusses the physical properties of graphene and its application such as in low-cost touchscreen, nanophonic devices, optoelectronics, photodiodes, and optical communication systems.

Section 2 contains Chapters 3, 4, and 5 dealing with photonic devices. Chapter 3 discusses studies on InAsP quantum dots to generate entangled photon generation and light extraction with a very high efficiency. Chapter 4 deals with the physics of interactions between nuclear particles such as positrons and electrons and hydrogenic targets and its applications. It presents the calculation of photodetachment and photoionization of two-electron systems useful to study the opacity of the sun. Chapter 5 describes the development of photovoltaic devices using perovskite materials like bismuth halide solar cells. The structures and their optoelectronic properties of the devices are reviewed and discussed.

Section 3 deals with the principle of semiconductor materials and devices. It contains Chapters 6, 7, and 8. Chapter 6 covers the principle and control of diffusion and intermixing of atoms in the semiconductor lattices. It summarizes the reported techniques to use the phenomena to create diffused quantum wells (QWs). Examples of QW semiconductor lasers and light-emitting diodes (LED) fabricated using inter-diffusion techniques are also given in Chapter 7. The scope of this chapter is to introduce a highly efficient HfO_2 atomic layer deposition (ALD) process with superior interface defect characteristics that can be applied on high-mobility III-V substrates. The effect of isopropyl alcohol precursor and in situ cyclic nitrogen plasma treatment on the HfO_2 ALD process at III-V substrates was thoroughly investigated. Chapter 8 discusses the basics of elastic and inelastic scattering. The inelastic scattering, Raman scattering phenomenon and surface-enhanced Raman scattering (SERS), are reviewed in detail. The chapter also discusses the latest developments in the SERS and its applications in various fields including food industries, detection of biomolecules, and in photovoltaic devices such as organic light emitting diodes.

Mojtaba Kahrizi and Parsoua Abedini Sohi
ECE Department,
Concordia University,
Montreal, Quebec, Canada

Section 1

Quantum Devices

Chapter 1

Principles and Applications of Nanoplasmonics in Biological and Chemical Sensing: A Review

Parsoua A. Sohi and Mojtaba Kahrizi

Abstract

Biosensing requires a highly sensitive real-time detection of the biomolecules. These properties are granted by nanoplasmonic sensing techniques. SPR-based optical sensors have evolved as a sensitive and versatile biosensing tool. A growing number of SPR-based sensing applications in the solution of clinical problems are reported in the recent years. This refers to the point that these sensors provide label-free detection of the living cells and non-destructive analysis techniques. In this study, we will review the mechanism of the detection in SPR biosensing, followed by the methods used to develop sensors to detect gases and the chemical, biological, and molecular interaction. The device sensitivity improvement based on plasmonic effects is also addressed in this study, and accordingly, the size and material dependence of the resonance frequency are discussed. The reviewed articles are categorized into three groups, depending on the SPR excitation configuration. In the first group of the sensors, the sensitivity of LSPR-based sensors in prism coupler configurations is reviewed. The second group, SPR excitation by optical fiber, slightly improved the sensitivity of the detections. The unique capability of the third group, photonic crystal fiber SPR sensors, in providing greatly improved sensitivity, generated a vast field of researches and applications in biosensing devices.

Keywords: surface plasmon resonance, localized surface plasmon resonance, nanoplasmonics, biosensors, extraordinary optical transmission, photonic crystals

1. Introduction

For the measurements of chemical and biological quantities in biomedical applications, a variety of methods were available and researched extensively in the last 30 years. In optical sensing, the first chemical sensors were based on the absorption spectrum of species to be measured [1]. Other chemical and biosensors have been developed since that time based on a diversity of optical techniques that involved luminescence, phosphorescence, fluorescence, particle light scattering, Raman scattering, ellipsometry, interferometry, and surface plasmon phenomena. For sensing purposes, measurements of the refractive index and absorbance or fluorescence properties of analyte molecules or a chemo-optical transducing medium were performed using those devices.

Among the different techniques developed, the potential of surface plasmon (SP) sensors was clearly recognized due to their great sensitivity, their simple structure, and their capacity in terms of real-time analysis of biospecific interactions without the use of labeled molecules. In fact, SP sensors have been the preferred devices used in real-time analysis situations. They are based on surface plasmon resonance, or collective electron oscillations, that may exist at a metal-dielectric interface. Light is used for excitation of the SP waves, resulting in the transfer of energy into the SP wave. As there is a strong concentration of the electromagnetic field in the dielectric, the propagation constant of the SP wave depends strongly on variations of the optical properties of the dielectric medium surrounding the metallic surface.

Light incident on a smooth metallic surface cannot excite surface plasmons due to the transverse character of the optical wave and the longitudinal character of the SP wave, preventing the coupling between the two. A coupling mechanism is thus necessary, and it can be provided by surface rugosity [2], a grating structure [3], or Kretschmann and Otto prism configurations [4, 5]. Based on this, we can classify the sensors according to the following: (a) sensors using optical prism couplers [4], (b) sensors using grating couplers [6], (c) sensors using optical fibers (SP resonance active metal layer deposited around the fiber core) [7, 8], and (d) sensors using integrated optical waveguides [9]. The sensor characteristics include (a) its sensitivity (derivative of the monitored SP resonance parameter, such as the resonant angle or the wavelength, with respect to the parameter to be determined, such as the refractive index or the overlayer thickness), (b) its resolution (minimum change in the parameter to be determined), and (c) its operating range (range of values of the parameter to be determined). As an example, resolution in the range of 4×10^{-5} refractive index unit (RIU) can be generally achieved in sensors using optical prism couplers, while sensors using grating couplers can demonstrate a resolution in the range of 10^{-6} RIU. Among existing sensing devices, SP sensors using optical fibers presented the best prospect for miniaturization, but it is no longer the case as can be seen below. For their part, sensors based on integrated optical waveguides are promising in the development of multichannel sensing devices, with however limitations on sensitivity performance due to the relatively low concentration of the electromagnetic field achieved in the analytc.

Recently, however, the event of ordered periodic nanostructures as well as of photonic crystal configurations has opened a wealth of new opportunities. Although nanoparticles are not new, new tools for their engineering in complex architectures at nanoscale have helped to discover and understand new and exciting phenomena, important not only for fundamental studies but also for device and system. Due to the renewed interest generated by the new fabrication tools (self-assembly methods, electron-beam lithographic methods, nanoimprint methods) and the extraordinary properties exhibited by ordered structures, a new term, "plasmonics," has been coined to describe the study of metallic and metallodielectric nanostructures and plasmon. The promise of highly integrated optical devices with structural elements smaller than the light wavelength made many to refer to it as the "next big thing" in nanotechnology.

2. Principle of operation

Surface plasmon is a charge density wave that exists at the interface between a metal and a dielectric. Surface plasmon polaritons (SPP) is collective oscillation of SPs, which is induced by an electromagnetic wave. This is the interesting optical property of noble metals, as these materials provide the best evidence of SPPs due to possession of high density of electrons free to move. SPs are optically excited at resonance condition. The resonance condition is referred to the condition where the momentum and energy of an incident photon matches that of a SP. Light energy can

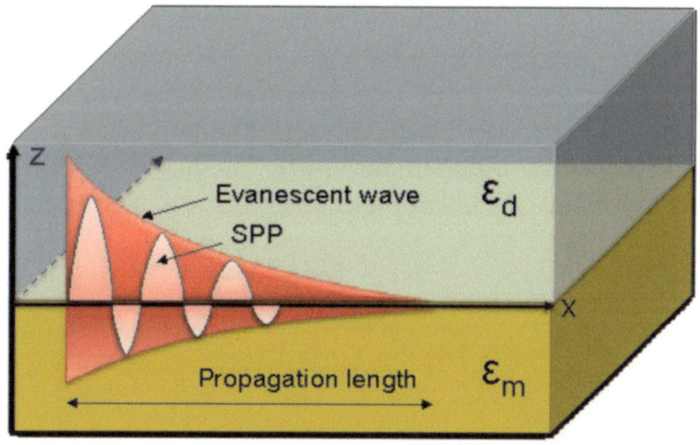

Figure 1.
SPR electromagnetic wave propagates parallel to the metal-dielectric interface. The evanescent field of SPPs is decaying in either side of the interface.

then be coupled to a SP wave which results in a strong absorption at a certain wavelength and angle of incidence. Accordingly there are two types of resonance measurement techniques: angular interrogations (measuring absorption as a function of incident angle) and wavelength interrogations (measuring absorption as a function of incident wavelength) [10, 11].

Surface plasmon resonance (SPR) can be seen as the electromagnetic surface waves that are the solution of Maxwell equation. According to the solution of Maxwell equation, SPR occurs at the interface of a material with a positive dielectric constant with that of a negative dielectric constant (such as noble metals). The waves propagate in the x-y plane along the metal-dielectric interface, and their lateral extensions evanescently decay into both sides of the interface (decay in the z-direction). The penetration length of the evanescent waves is longer for the dielectric medium compared to the metallic side (**Figure 1**).

S-polarized light (TE polarization) is referred to the polarization state that the electric field is parallel to the surface of the interface. P-polarized (TM polarized) light is the light that the electric field lays on the plane of the incident (the perpendicular plane to the surface of the interface and contains the wavevector of the excitation source). To excite the SPP modes, it is required to have the components of the electric field acting along the metal-dielectric interface. Hence, the oblique incident of the P-polarized light, which has the nonzero perpendicular electric vector component on the z-axis, is required.

P-polarized light with a particular wavelength and incident angle induces SPR once the wavevector value of the SP wave is identical to that of the incident light. The dispersion relation of the SPPs can be derived from the solution of wave equation for a P-polarized electromagnetic incident light governed by Maxwell equations subjected to the continuity of the tangential and normal field component as the boundary conditions. The wavevector of the SPP (K_{SPP}) is determined by the following relation:

$$K_{SPP} = \frac{\omega}{c}\sqrt{\frac{\varepsilon_m \varepsilon_d}{\varepsilon_m + \varepsilon_d}} \tag{1}$$

where ω is the frequency of excitation source, c is velocity of the light, and ε_m and ε_d are dielectric constants of the metal and dielectric material, respectively [12]. As the real part of the K_{SPP} for noble metals always falls below the wavevector of the incident light $\left(K_x = \frac{2\pi}{\lambda} sin\theta\right)$, to match the wave vectors, specific excitation configuration is required. The prism coupling is a simple and standard technique to induce SPR. The underlying principle of this is to bring down the wavevector of the incident light by passing through a high refractive index prism with $n_{prism} > n_{dielectric}$ in a specific incident angle (θ). The wavevector matching at resonance condition for the prism coupler configuration is defined by

$$K_x = K_{SPP} \tag{2}$$

$$\frac{2\pi}{\lambda} n_{prism} \sin\theta = \frac{2\pi}{\lambda} Re\left\{\sqrt{\frac{\varepsilon_m \varepsilon_d}{\varepsilon_m + \varepsilon_d}}\right\} \tag{3}$$

where λ is the wavelength of the incident light [10]. Equation (3) states that as the refractive index of the dielectric is altered, the propagation constant of the SP mode is altered. This results in changing the coupling conditions between the incident light and SP modes.

Figure 2 demonstrates the SPR experimental setup using Kretschmann prism coupling [13, 14], which provides proper synchronization between K_{SPP} and K_x. **Figure 2b** shows the dispersion relation for SP in matter. It is seen that by using the prism, it is possible to couple the wavevectors of the SP with that of the incident light.

There are other SPR excitation configurations reported in the literature such as Otto prism coupler [15], waveguide and fiber-optic coupler [16], and diffraction grating [6]. The principal of the waveguide (and fiber-optics) coupler is close to what was explained for prism coupler, except in this configuration it is not possible to interrogate the incident angle (**Figure 3a**). In diffraction grating configuration, the incident electromagnetic radiation is directed toward a medium whose surface has a spatial periodicity (Λ) similar to the wavelength of the radiation. The incident beam is diffracted in different orders producing propagating evanescent modes at the interface. The evanescent modes have wavevectors parallel to the interface similar to the incident radiation but with integer "quanta" of the grating wavevector added or subtracted from it. These modes couple to SP, which run along the interface between the grating and the ambient medium. The diffraction grating is the only configuration that generates the SP in the same side of the metallic layer as the incident light. The only drawback of this configuration is having more than one peak in the resonance spectrum as there are several orders of diffractions in the reflected beam [17].

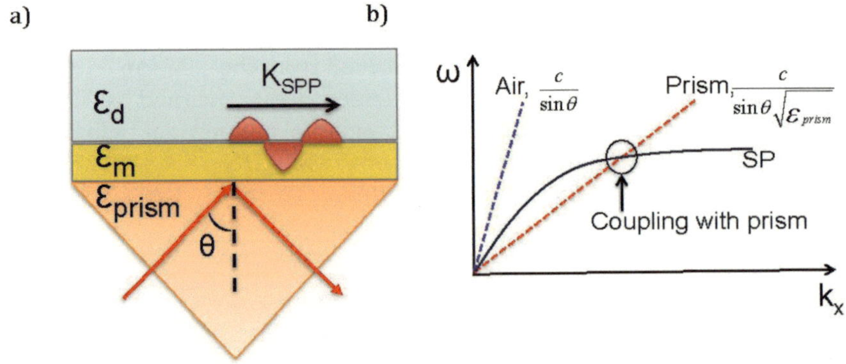

Figure 2.
(a) Prism coupling, Kretschmann configuration, and (b) dispersion relation of incident light coupling SP.

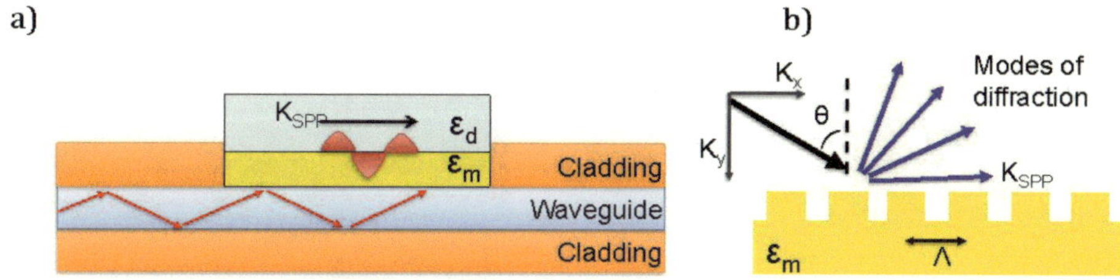

Figure 3.
(a) Waveguide coupler and (b) diffraction grating.

In the last many years, numerous researches have involved coupling photons to surface plasmons to study the effect of various materials, structure, and configurations on resonance spectrum. Potential applications extend to new light sources, solar cells, holography, Raman spectroscopy, microscopy, and sensors. In the following sections, the recent advances of SPR sensing applications are explored.

3. Applications in biological and chemical sensing

Numerous useful strategies of protein labeling have been developed for biophysical characterization of proteins including their structure, folding, and interaction with other proteins [18]. Molecular labels, such as biotin, reporter enzymes, fluorophores, and radioactive isotopes, attach covalently to target protein and nucleotides to facilitate identifying and quantifying of labeled target [19]. As the use of the molecular labels may modify the structural configuration and the binding properties of the molecules of interest, a variety of label-free methods have been developed. Among the various sensing methodologies, the SPR-based system is a reliable type of label-free technique for monitoring biomolecular interactions particularly in thermodynamic and kinetic analyses [18–21]. Biosensing application of SPR was first reported in 1983 [22, 23]. As it was mentioned in Section 2, SPR sensors are sensitive to changes in refractive index of the bulk solution in the vicinity of their active surface. Although several techniques are proposed to improve the sensitivity of SPR measurements [5, 24], the sensitivity to monolayers or molecular binding is obtained by the confinement of the plasmonic field in the nanostructured noble metals, which is known as LSPR. The smaller decay length associated with LSPR than that of SPR makes it more sensitive to local refractive index surrounding the nanostructures. **Figure 4a** illustrates the sensitivity of the LSPR spectrum of the functionalized gold nanostructures (shown in **Figure 4b**) toward adsorption of amyloid β-derived diffusible ligands (ADDL).

The impact of the shape and size of the nanostructured noble metals on the device sensitivity is demonstrated by many research groups. Analytical studies revealed that the sharp-tipped structures are reinforcing higher refractive index sensitivity due to the higher electromagnetic fields produced on their sharp edges. Reportedly, particles with sharp tips produce much higher refractive index sensitivities that would be predicted from their aspect ratios alone [12, 26–28].

Several geometries have been widely explored in recent years for developing LSPR sensors. Ekinci et al. [29] compared the extinctions spectra of Au, Al, and Ag nanoparticles of the same size. Well-defined ordered nanoparticle arrays were fabricated on quartz substrates. The diameter of the structures varied between 20 and 140 nm. Al nanoparticles with 40 nm diameters exhibit strong and sharp plasmon resonances in the near- and deep-UV ranges. Peaks in visible range were

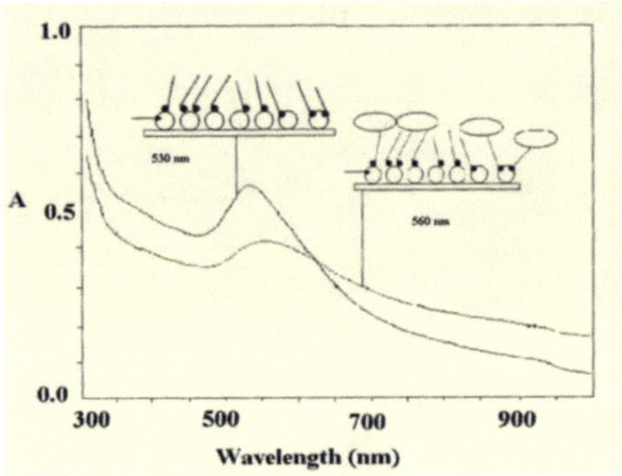

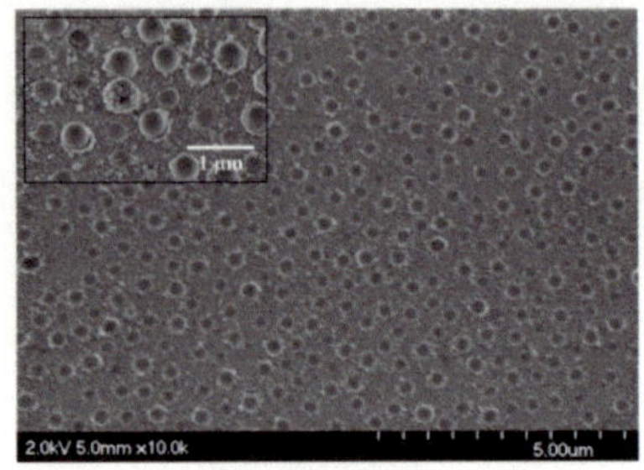

Figure 4.
(a) UV-VIS spectrum of ADDLs on gold and (b) nanohole/nanoring array prepared with 530 nm polystyrene (PSo microspheres) and 20 nm Au layer. The inset shows enlarged image of a region where PS spheres were not completely removed. Au nanoparticles are around and on the top of the spheres [25].

observed for Ag and Au nanoparticles. Increasing the diameter of the nanoparticles for tested materials resulted in a red shift in the peak position.

Considering the feasibility of fabrication, nanoring and nanodisk structures have gained great interest for their biological and chemical sensing properties. Larsson et al. [30] compared the LSPR resonance sensitivity of gold nanorings and nanodisks to refractive index changes of bulk and thin dielectric films with different thicknesses. The sensitivity of nanoring structures was found significantly higher than that of nanodisks with similar diameters in the near-infrared spectrum. Tuning of the LSPR peak wavelength between approximately 1000–1300 nm was obtained by changing the diameter of the nanorings. It is also reported that the extinction spectrum [31] of the nanoring structure is affected by the angle of incidence. By varying the incidence angle from normal to oblique, several higher-energy plasmon resonances appear in the extinction spectrum of the nanoring structures [32].

The mechanism of the enhancement of plasmon resonance in a ring array is analytically studied by Wang et al. [33]. They successfully studied the sensitivity of the device based on the dimension and periodicity of the nanorings. According to their results, increasing the thickness of the nanorings introduces a blue shift in resonance peak wavelength. For the thicknesses above 40 nm, the second resonant peak appears near the short wavelength (below 600 nm). Regarding the periodicity, as the period decreases, a red shift was observed in the position of resonance peak. Sensing characteristics of the optimum sensor was experimentally tested for refractive indices in the range of 1.33–1.4 (obtained from different ratios of glycerol water mixtures). The results also showed a linear relationship of the peak position and refractive index of the medium.

Although biological sensing applications of nanostructure-based LSPR have been reported in several laboratory level studies, commercialized implementation of this technique still requires large improvement regarding reproducibility of the structures in terms of size and shape of nanoparticles. Reproducible structures such as patterned thin films either with perforations or protrusions show a similar plasmonic response [34].

Parsons et al. [35] compared the plasmonic response of nanoparticle and nanohole arrays. They have shown that the spectral response depends on inter-hole separation, while there seems to be little effect of the interparticle spacing. This conclusion is referred to the coupling mechanism of the SPP mode that the thin metal film supports. It was also shown that the resonant spectrum of the thin film perforated with nanohole arrays is qualitatively similar to a particle of approximately the same dimensions and quantitatively shows weaker local field enhancements. Conclusively, in general, LSPR of nanoparticles is more suitable for applications which rely on large local enhancements of electric field, such as surface enhanced Raman scattering.

Another form of SPR-based sensors involves extraordinary light transmission (EOT). Periodic structures consisting of a thin metallic film perforated with an array of nanoscale holes exhibit EOT provided that the hole size is in the subwavelength range. This porous structure can convert light into SPs by providing the necessary momentum conservation for the coupling process. EOT is determined when the transmission spectrum contains a set of peaks with enhanced transmission, although the individual holes are so small that they do not allow propagation of light. Multiple EOT peaks are correlated with excitation of various SPP modes. These peaks are generated by excitation of LSP modes which occurs in the inner

Structure	Study*	Sensitivity (nm/RIU)	Range of λ	Medium/RI	Comment	Ref
Au nanoring	Experimental	880	NIR	1.33–1.4	LSPR. Effect of size is studied	[30]
Double split Au nanoring	Analytical	1200	NIR	1.33–1.38	LSPR. Effect of ring configuration is studied	[43]
Au nanowell	Experimental	1200–1600	Vis-NIR	1–1.6	LSPR. Tuned based on geometrical parameters	[44]
Au thin film with cubic nanoholes	Analytical	2000	Vis	1.333 and 1.357	LSPR. In this study protrusive thin films showed higher sensitivity	[10]
Au thin film with circular nanoholes	Experimental	300	Vis	1.333–1.35	LSPR. Dielectric spacer layer between plasmonic film and substrate improved the sensitivity	[45]
Au thin film with circular nanoholes	Experimental	200	Vis-NIR	1–1.5	LSPR. The resonance peak was tuned from 650 to 850 nm by varying the interhole separation	[46]
Au nanoparticle	Experimental	57	UV-Vis	1–1.658	LSPR. Nonuniform structures. Particles with mean diameters of 18 ± 2 nm	[47]
Au nanoring	Experimental	—	UV-Vis	ADDL fibrinogen AT5G0701	LSPR	[25]
Quantum dots of copper sulfate	Experimental	—	NIR	1.46, 1.51, 1.63	LSPR, refractive indices refer to CCl4, TCE, and CS2	[48]
Au colloidal nanoparticles	Experimental	70	Vis	1.32–1.5	LSPR, 30 nm diameter of the particles	[49]
Au thin film with nanoholes	Experimental	481	Vis-NIR	1.33–1.36	EOT, holes of 200 nm diameter with 500 nm periodicity	[50]
Au nanorods	Experimental	650	NIR	1.34–1.7	LSPR, tunable with size of the particles	[51]

**In most reported researches, analytically studied sensors show higher sensitivities than that of the experimental studies.*

Table 1.
Characteristics of several reported SPR-based biological/chemical sensors.

surface of hole arrays and SPP modes at the surface of the thin film, in the upper and lower rims of the holes [36].

The transmission of electromagnetic waves through a subwavelength hole was investigated by Bethe [37] for the first time. The transmission will occur at specific frequencies imposed by geometrical parameters of the structure, polarization, angle of incident light, and permittivity of the surrounding media. These findings found many applications owing to the simplicity with which their spectral properties can be tuned on.

In several studies [38–41] it was shown that the peak positions are determined by the periodicity of the holes. The periodicity is usually comparable to the wavelength of the incident light. By modifying the film thickness and diameter of the holes, one can control the shape of the peaks in terms of the width and intensity. It was due to these facts that many researchers believe that SPP is responsible for EOT and developed models to explain the phenomena.

The simple structure (perforated thin films) of EOT-based devices is widely studied for potential applications, from optical switches and photolithography masks to sensing applications. Despite this simple structure, the quality of the noble metal film in terms of surface roughness, grain size, and purity is equally important for EOT device application. A single crystalline thin film with atomically smooth surface is required in order to maximize the propagation length of SPP. A smooth surface results in elimination of scattering and loss of the incident light. This improves the transmission spectrum of the device, though it is very challenging to synthesize or grow uniform single crystalline films over a large area [42].

Table 1 summarizes the reported characteristics of few SPR-, LSPR-, and EOT-supported sensors. As it can be seen, the geometry of the metal film plays an important role in plasmon frequency. As an example, gold has plasmon frequency in the deep ultraviolet; however geometric factors create the possibility to tune the resonance peak wavelength range. The sensitivity (S) of the sensors utilizing the wavelength interrogation is defined as

$$S\left(\frac{nm}{RIU}\right) = \frac{\delta\lambda_{res}}{\delta n_d} \tag{4}$$

where $\delta\lambda_{res}$ is the offset of the resonance peak and δn_d is the change in the refractive index of the dielectric medium.

4. SPR sensitivity improvements using photonic crystal fiber

As mentioned in previous sections, the conventional prism-based Kretschmann setup is widely used for SPR sensors. However, this configuration is bulky due to the required optical measurement components. Furthermore, this configuration is not adequate for remote sensing [7]. The remarked limitations of conventional SPR configurations led to the use of optical fiber instead of the prism. Chemical sensing application of optical fiber configuration was proposed by Jorgenson et al. [52, 53] for the first time, and since then the advantages of various configurations of optical fiber based SPR sensors have drawn a lot of attention. Sensing properties of numerous microstructure optical fiber (MOF)-based SPR have been reported in the last two decades. The sensing properties of some of these sensors are summarized in **Table 2**.

Further improvement in terms of sensitivity and resolution is obtained by photonic crystal fiber-based SPR (PCF SPR). PCFs are similar to conventional optical fiber, but with periodic air holes inside the cladding region. The diameter of the air

Structure	Study	Sensitivity (nm/RIU)	Range of λ	Medium/ RI	Comment	Ref
Single mode, Au-coated core	Experimental	3200	Vis	1.33–1.3385	The sensor is able to detect index changes as low as 4×10^{-6} under moderate fiber deformations	[57]
Tapered fiber, Au-coated	Experimental	400	Vis-NIR	1.35–1.42	The sensor provides qualitative and quantitative biochemical detection	[58]
Biconical tapered, Au-coated	Experimental	2100	Vis	1.326–1.375	—	[59]
Uncladded fiber, Au-coated	Experimental	1700	Vis-NIR	1.333–1.35	Red shift was observed in resonance wavelength of the patterned Au layer	[60]

Table 2.
Characteristics of several reported fiber-optic SPR-based sensors.

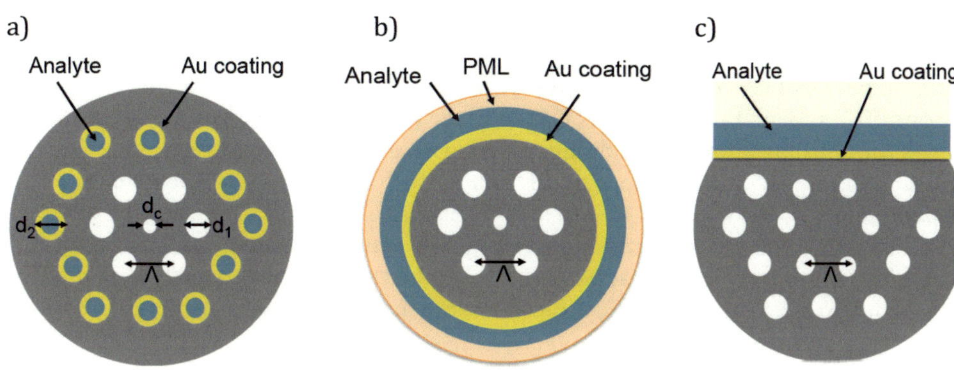

Figure 5.
Schematics of PCF-SPR sensors. (a) Analyte-filled cladding holes [54], (b) externally coated PCF [55], and (c) D-shaped PCF SPR sensor [56].

Structure	Study	Sensitivity (nm/RIU)	Range of λ	Medium/ RI	Comment	Ref
Gold-coated D-shaped PCF	Experimental and analytical	46,000	Vis-NIR	1.33–1.42	This is the maximum sensitivity. The average sensitivity of 9800 nm/RIU is reported	[54]
Silver-graphene, D-shaped PCF	Analytical	3700	Vis	1.33–1.37	This sensitivity is found for RI of 1.36	[61]
Two parallel Au-coated D-shaped PCFs	Analytical	13,500	NIR	1.27–1.33	13,500 is the maximum sensitivity for RI of 1.32	[62]
PCF externally coated with TiO_2-Au	Analytical	23,000	Vis-NIR	1.32–1.40	23,000 is the maximum sensitivity for RI of 1.32	[63]

Structure	Study	Sensitivity (nm/RIU)	Range of λ	Medium/ RI	Comment	Ref
PCF with Au-metalized microfluidic slots	Analytical	2000	Vis	1.33–1.34	The effect of geometrical parameters is studied. Sensitivity is inversely proportional to gold layer thickness	[64]
PCF with Au-coated multichannel	Analytical	2400	Vis	1.33–1.34	The effect of geometrical parameters is studied	[65]
Analyte filled with Au-coated core PCF	Analytical	2280	NIR	1.46–1.485	Several peaks were observed in the resonance spectrum	[66]

Table 3.
Characteristics of several reported PCF SPR sensors.

holes and the separation gap between them defines the light propagation characteristics of PCFs. These sensors own other advantages including small size and flexible structural design over conventional optical fiber and prism coupler SPR sensors. To date, numerous PCF SPR sensors have been demonstrated with different configurations. The schematics of some configurations are shown in **Figure 5**. **Figure 5a** shows the configuration in which the metal layer and the analyte are filled inside the cladding holes. External coatings and D shape PCF are shown in **Figure 5b** and **c** respectively.

The performance of recently reported PCF-based SPR sensors are compared in **Table 3**. Most of the reported sensors are based on theoretical and analytical studies. This is due to the fact that fabrication of these sensors requires a complex process.

5. Conclusion

In this review, the basic principles of SPR and various configurations of SPR excitation are reviewed. Since nanostructures of noble metals have tunable optical properties, they can be applied in nanoplasmonics-based devices and sensors. Nanoplasmonics has proven to be useful in sensing applications, especially for biological uses. Compared to SPR-based sensors, LSPR-based sensors exhibit unique properties, including higher sensitivities and figure of merits (FOM).

The researchers' aim to reduce the size of SPR- and LSPR-based sensors led to development of fiber-optic SPR-based sensors. The highest sensitivity reported in the literature allocates to PCF SPR sensors. They have shown their capability in providing high sensitivity as high as 46,000 nm/RIU [54] with respect to small RI changes. However, the majority of reported PCF SPR sensors are based on analytical studies (as it is shown in **Table 3**, most of the researches are only carried out with numerical simulations). This is due to complications of the fabrication process of the PCF SPR sensor. The complication mostly refers to interior metallic coating (into the core or cladding holes). Large interest in PCF SPR sensors among researchers increases the importance of the fabrication technique improvements.

Author details

Parsoua A. Sohi and Mojtaba Kahrizi*
Department of Electrical and Computer Engineering, Concordia University, Montreal, Quebec, Canada

*Address all correspondence to: mojtaba.kahrizi@concordia.ca

IntechOpen

© 2020 The Author(s). Licensee IntechOpen. This chapter is distributed under the terms of the Creative Commons Attribution License (http://creativecommons.org/licenses/by/3.0), which permits unrestricted use, distribution, and reproduction in any medium, provided the original work is properly cited.

References

[1] Lubbers DW, Opitz N. The pCO_2-/pO_2-optode: A new probe for measurement of pCO_2 or pO_2 in fluids and gases. Journal of Biosciences. 1975; **30**(7–8):532-533

[2] Kanso M, Cuenot S, Louarn G. Roughness effect on the SPR measurements for an optical fibre configuration: Experimental and. Journal of Optics A: Pure and Applied Optics. 2007;**9**:586-592

[3] SEO M, LEE J, LEE M. Grating-coupled surface plasmon resonance on bulk stainless steel. Optics Express. 2017;**25**(22):254-262

[4] Svedendahl M, Chen S, Kall M. An Introduction to Plasmonic Refractive Index Sensing. In: Dmitriev A, editor. Nanoplasmonic Sensors. Springer, NY: Integrated Analytical Systems. 2012. Available from: https://link.springer.com/chapter/10.1007/978-1-4614-3933-2_1

[5] Akowuah EK, Gorman T, Haxha S. Design and optimization of a novel surface plasmon resonance biosensor based on Otto configuration. Optics Express. 2009;**17**(26):491-496

[6] Rossi S, Gazzola E, Capaldo P, Borile G, Romanato F. Grating-coupled surface plasmon resonance (GC-SPR) optimization for phase-interrogation biosensing in a microfluidic chamber. Sensors. 2018;**18**(1621):1-13

[7] Gupta BD, Verma RK. Surface plasmon resonance-based fiber optic sensors: Principle, probe designs, and some applications. Journal of Sensors. 2009:1-12. DOI: 10.1155/2009/979761

[8] Zhang C, Li Z, Zhen S, Hui C, Cai S, Yu J. U-bent fiber optic SPR sensor based on graphene/AgNPs. Sensors and Actuators B: Chemical. 2017;**251**:127-133

[9] Ctyroky J et al. Theory and modelling of optical waveguide sensors utilising surface plasmon resonance. Sensors and Actuators B: Chemical. 1999;**54**(1–2): 66-73

[10] Dormeny AA, Sohi PA, Kahrizi M. Design and simulation of a refractive index sensor based on SPR and LSPR using gold nanostructures. Results in Physics. 2020;**16**:102869

[11] Fida F. Biosensing Based on Localized Surface Plasmon Resonance of Gold Nanostructures Fabricated by a Novel Nanosphere Lithography Technique. [Thesis]. Concordia University; 2008

[12] Prabowo BA, Purwidyantri A, Liu K-C. Surface plasmon resonance optical sensor: A review on light source technology. Biosensors. 2018;**8**(80):1-27

[13] Kretschmann E, Raether H. Radiative decay of non-radiative surface plasmon excited by light. Zeitschrift für Naturforschung A. 1968;**23**:2135-2136

[14] Roh S, Chung T, Lee B. Overview of the characteristics of micro- and nano-structured surface plasmon resonance sensors. Sensors. 2011;**11**(2):1565-1588

[15] Otto A. Excitation of nonradiative surface plasma waves in silver by the method of frustrated total reflection. Zeitschrift für Physik. 1968;**216**:398-410

[16] Allsop T, Neal R. A review: Evolution and diversity of optical fibre plasmonic sensors. Sensors. 2019; **19**(4874):1-19

[17] Petty CM. Molecular Electronics: From Principle to Practice. Wiley: United States; 2008

[18] Nguyen HH, Park J, Kang S, Kim M. Surface plasmon resonance: A versatile

technique for biosensor applications. Sensors. 2015;**15**:10481-10510

[19] Syahir A, Usui K, Tomizaki K, Kajikawa K, Mihara H. Label and label-free detection techniques for protein microarrays. Microarrays. 2015;**4**(2): 228-244

[20] Vollmer F, Arnold S. Whispering-gallery-mode biosensing: Label-free detection down to single molecules. Nature Methods. 2008;**5**(7):591-596

[21] Méjard R, Griesser HJ, Thierry B. Optical biosensing for label-free cellular studies. Trends in Analytical Chemistry. 2014;**53**:178-186

[22] Liedberg B, Nylander C, Lundstrm I. Biosensing with surface plasmon resonance—How it all started. Biosensors & Bioelectronics. 1995;**10**:1-9

[23] Liedberg B, Nylander C, Lunstrom I. Surface plasmon resonance for gas detection and biosensing. Sensors and Actuators. 1983;**4**:299-304

[24] Sathiyamoorthy K, Ramya B, Murukeshan VM, Wei Sun X. Modified two prism SPR sensor configurations to improve the sensitivity of measurement. Sensors and Actuators A: Physical. 2013; **191**:73-77

[25] Fida F, Varin L, Badilescu S, Kahrizi M. Gold nanoparticle ring and hole structures for sensing proteins and antigen–antibody interactions. Plasmonics. 2009;**4**:201-207

[26] Mock JJ, Hill RT, Tsai Y, Chilkoti A, Smith DR. Probing dynamically tunable localized surface plasmon resonances of film-coupled nanoparticles by evanescent wave excitation. Nano Letters. 2012;**12**:1757-1764

[27] Petryayeva E, Krull UJ. Localized surface plasmon resonance: Nanostructures, bioassays and biosensing—A review. Analytica Chimica Acta. 2011;**706**(1):8-24

[28] Tosi D, Poeggel S, Iordachita I, Schena E. Fiber optic sensors for biomedical applications. Opto-Mechanical Fiber Optic Sensors. 2018: 301-333. DOI: 10.1126/science.6422554

[29] Ekinci Y, Solak HH, Löffler JF. Plasmon resonances of aluminum nanoparticles and nanorods. Journal of Applied Physics. 2008;**104**(083107): 1-6

[30] Larsson EM, Alegret J, Ka M, Sutherland DS. Sensing characteristics of NIR localized surface plasmon resonances in gold nanorings for application as ultrasensitive biosensors. Nano Letters. 2007;**7**(5):1256-1263

[31] Movsesyan AR, Baudrion A-L, Adam P-M. Extinction measurements of metallic nanoparticles arrays as a way to explore the single nanoparticle plasmon resonances. Optics Express. 2018;**26**(5): 2949-2951

[32] Hao F, Larsson EM, Ali TA, Sutherland DS, Nordlander P. Shedding light on dark plasmons in gold nanorings. Chemical Physics Letters. 2008;**458**(4–6):262-266

[33] Wang S, Sun X, Ding M, Peng G. The investigation of an LSPR refractive index sensor based on periodic gold nanorings array. Journal of Physics D: Applied Physics. 2018;**51**(4):045101 (7pp)

[34] Canpean V, Astilean S. Multifunctional plasmonic sensors on low-cost subwavelength metallic nanoholes arrays. Royal Society of Chemistry. 2009;**9**(24):3574-3579

[35] Parsons J, Hendry E, Burrows CP, Auguié B, Sambles JR, Barnes WL. Localized surface-plasmon resonances in periodic nondiffracting metallic

nanoparticle and nanohole arrays. Physical Review B. 2009;**79**(7):1-4

[36] Chen Z, Li P, Zhang S, Chen Y, Liu P. Enhanced extraordinary optical transmission and refractive-index sensing sensitivity in tapered plasmonic nanohole arrays. Nanotechnology. 2019; **30**:335201(9pp)

[37] Bethe H. Theory of diffraction by small holes. Physical Review. 1944;**66** (7–8):163

[38] Hajiaboli A. Optical properties of thick metal nanohole arrays fabricated by electron-beam and nanosphere lithography. Physica Status Solidi A: Applications and Material Science. 2009;**206**(6668):976-979

[39] Rodrigo SG, de León-Pérez F, Martín-Moreno L. Extraordinary optical transmission: Fundamentals and applications. Proceedings of the IEEE. 2016;**104**(12):2288-2306

[40] Motogaito A, Morishita Y, Miyake H, Hiramatsu K. Extraordinary optical transmission exhibited by surface plasmon polaritons in a double-layer wire grid polarizer. Plasmonics. 2015;**10**(6):1657-1662

[41] Liu H, Lalanne P. Theory of the extraordinary optical transmission. Nature Letters. 2008;**452**:728-731

[42] Zhang J, Irannejad M, Yavuz M, Cui B. Gold nanohole array with sub-1 nm roughness by annealing for sensitivity enhancement of extraordinary optical transmission biosensor. Nanoscale Research Letters. 2015;**10**(238):1-8

[43] Liu S, Yang Z, Liu R, Li X. High sensitivity localized surface plasmon resonance sensing using a double split nanoring cavity. The Journal of Physical Chemistry C. 2011;**115**(50): 24469-24477

[44] Lee SY, Kim S-H, Jang SG, Heo CJ, Shim JW, Yang SM. High-fidelity optofluidic on-chip sensors using well-defined gold nanowell crystals. Analytical Chemistry. 2011;**83**(23): 9174-9180

[45] Bochenkov VE, Frederiksen M, Sutherland DS. Enhanced refractive index sensitivity of elevated short-range ordered nanohole arrays in optically thin plasmonic Au films. Optics Express. 2013;**21**(12):4428-4433

[46] Brian B, Sepúlveda B, Alaverdyan Y, Lechuga LM, Käll M. Sensitivity enhancement of nanoplasmonic sensors in low refractive index substrates. Optics Express. 2015; **17**(3):2335-2339

[47] Toderas F, Baia M, Baia L, Astilean S. Controlling gold nanoparticle assemblies for efficient surface-enhanced Raman scattering and localized surface plasmon resonance sensors. Nanotechnology. 2007;**18**: 255702

[48] Luther JM, Jain PK, Ewers T, Alivisatos AP. Localized surface plasmon resonances arising from free carriers in doped quantum dots. Nature Materials. 2011;**10**(4):361-366

[49] Sun Y, Xia Y. Increased sensitivity of surface plasmon resonance of gold nanoshells compared to that of gold solid colloids in response to environmental changes. Analytical Chemistry. 2002;**74**(20):5297-5305

[50] Im H, Sutherland JN, Maynard JA, Oh S. Nanohole-based surface plasmon resonance instruments with improved spectral resolution quantify a broad range of antibody-ligand binding kinetics. Analytical Chemistry. 2012;**84**: 1941-1947

[51] Lee KS, El-Sayed MA. Gold and silver nanoparticles in sensing and imaging: Sensitivity of plasmon

response to size, shape, and metal composition. Journal of Physical Chemistry B. 2006;**110**(39):19220-19225

[52] Jorgenson R, Yee S, Johnson K, Compton B. Novel surface-plasmon-resonance-based fiber optic sensor applied to biochemical sensing. In: Fiber Optics Sensors in Medical Diagonistics. 1993. pp. 35-48. DOI: 10.1117/12.144841

[53] Jorgenson R, Yee S. A fiber-optic chemical sensor based on surface plasmon resonance. Sensors and Actuators B: Chemical. 1993;**12**(3): 213-220

[54] Rifat AA, Ahmed R, Mahdiraji GA, Adikan FRM. Highly sensitive D-shaped photonic crystal fiber-based plasmonic biosensor in visible to near-IR. IEEE Sensors Journal. 2017;**17**(9):2776-2783

[55] Liu C et al. Numerical analysis of a photonic crystal fiber based on a surface plasmon resonance sensor with an annular analyte channel. Optics Communications. 2017;**382**:162-166

[56] Gangwar RK, Singh VK. Highly sensitive surface plasmon resonance based D-shaped photonic crystal fiber refractive index sensor. Plasmonics. 2017;**12**:1367-1372

[57] Piliarik M, Homola J. Surface plasmon resonance sensor based on a single-mode polarization-maintaining optical fiber. Sensors and Actuators B: Chemical. 2003;**90**:236-242

[58] Lin H-Y, Huang C-H, Cheng G-L, Chen N-K, Chui H-C. Tapered optical fiber sensor based on localized surface plasmon resonance. Optics Express. 2012;**20**(19):21693-21701

[59] Kim Y-C, Peng W, Banerji S, Booksh KS. Tapered fiber optic surface plasmon resonance sensor for analyses of vapor and liquid phases. Optics Letters. 2005;**30**(17):2218-2220

[60] Antohe I, Spasic D, Delport F, Li J, Lammertyn J. Nanoscale patterning of gold-coated optical fibers for improved plasmonic sensing. Nanotechnology. 2017;**28**:215301

[61] Dash JN, Jha R. On the performance of graphene-based D-shaped photonic crystal fibre biosensor using surface plasmon resonance. Plasmonics. 2015; **10**:1123-1131

[62] Wang F, Liu C, Sun Z, Sun T, Liu B. A highly sensitive SPR sensors based on two parallel PCFs for low refractive index detection. IEEE Photonics Journal. 2018;**10**(4):1-10

[63] Al Mahfuz M, Hossain MA, Haque E, Hai NH, Namihira Y, Ahmed F. A bimetallic-coated, low propagation loss, photonic crystal fiber based plasmonic refractive index sensor. Sensors. 2019;**19**(3794):1-12

[64] Akowuah EK et al. Numerical analysis of a photonic crystal fiber for biosensing applications. IEEE Journal of Quantum Electronics. 2012;**48**(11): 1403-1410

[65] Azzam SI, Hameed MFO, Shehata R, Heikal A, Obayya SS. Multichannel photonic crystal fiber surface plasmon resonance based sensor. Optical and Quantum Electronics. 2016;**48**(2):1-11

[66] Qin W, Li S, Yao Y, Xin X, Xue J. Analyte-filled core self-calibration microstructured optical fiber based plasmonic sensor for detecting high refractive index aqueous analyte. Optics and Lasers in Engineering. 2014;**58**:1-8

Chapter 2

Graphene-Based Nanophotonic Devices

Ankur Pandya, Vishal Sorathiya and Sunil Lavadiya

Abstract

Graphene is an ideal 2D material that breaks the fundamental properties of size and speed limits by photonics and electronics, respectively. Graphene is also an ideal material for bridging electronic and photonic devices. Graphene offers several functions of modulation, emission, signal transmission, and detection of wideband and short band infrared frequency spectrum. Graphene has improved human life in multiple ways of low-cost display devices and touchscreen structures, energy harvesting devices (solar cells), optical communication components (modulator, polarizer, detector, laser generation). There is numerous literature is available on graphene synthesis, properties, devices, and applications. However, the main interest among the scientist, researchers, and students to start with the numerical and computational process for the graphene-based nanophotonic devices. This chapter also includes the examples of graphene applications in optoelectronics devices, P-N junction diodes, photodiode structure which are fundamental devices for the solar cell and the optical modulation.

Keywords: Graphene, PN diode, photodetector, modulator, transistor

1. Introduction

The scientific community across the globe considered graphene as one of the most revolutionary 2D nanomaterials of the world that possesses zero energy bandgap [1]. As shown in **Figure 1**, being the monoatomic carbon layer arranged in a honeycomb lattice, graphene possesses unique attractive properties and by the virtue of which it has attracted strong scientific and technological interests [2–5]. Graphene has shown great application potential in many fields, such as nanoelectronics [5, 6], energy storage devices [6–10] and bioelectronic device applications [11–13]. The linear dispersion relation in graphene at the ends of the First Brillouin Zone is shown in **Figure 2** according to which the bandgap at the Dirac point is zero. However, zero energy bandgap limits the applications of pristine graphene on a wide-scale especially in graphene-based nanoelectronics because, in the absence of energy bandgap, it is not possible for pristine graphene-based electronic devices to be operated in ON and OFF states which is desirable for logic gate circuits. To overcome such limitations of pristine graphene and to improve its applicability for designing nanoelectronic devices, the interest also builds up towards the study on doped graphene. Suitable doping with proper concentration can introduce the desired bandgap in graphene that enables graphene to be utilized in electronic circuits at the nanoscale. It has been reported that the hydrogen passivated armchair GNR exhibits direct bandgap at the edges [14]. Moreover, the bandgap can be introduced and tuned by transforming

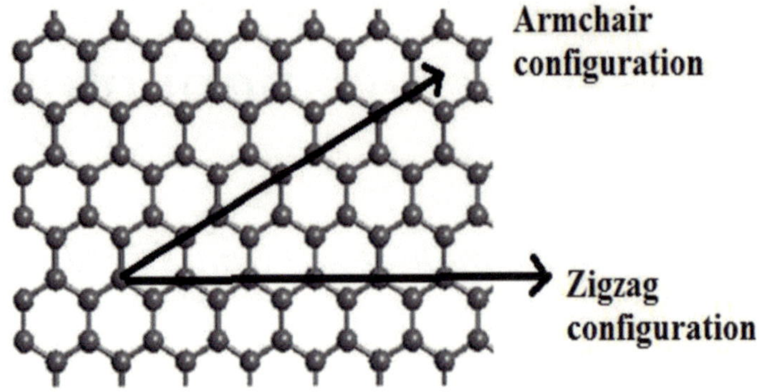

Figure 1.
Monolayer graphene nanoribbon [17].

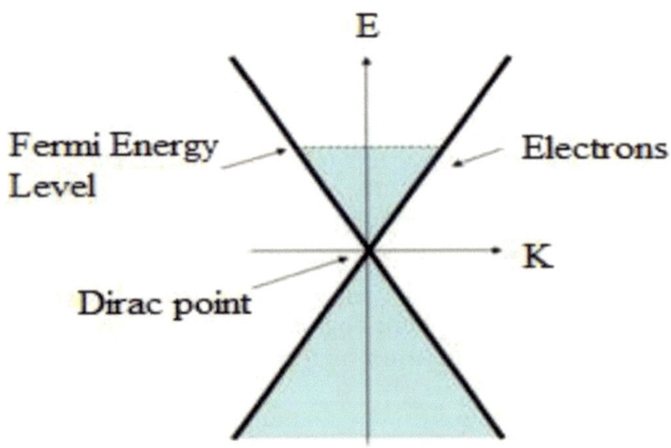

Figure 2.
The linear dispersion relation in graphene at the ends of First Brillouin Zone [18].

graphene sheet into its nanoribbon form of finite width i.e. graphene nanoribbon (GNR) [15, 16]. However, the bandgap of graphene nanoribbons shows different magnitudes for three groups i.e. Na =3p, 3p + 1, and 3p + 2 where Na is the number of dimers and p is an integer [15]. The magnitude of bandgap oscillates between these three groups with the number of dimers (N) i.e. $E_{g(3p+1)} > E_{g(3p)} > E_{g(3p+2)}$. The energy band gap of armchair GNR is determined as a function of the number of dimers wherein the smaller the number of dimer lines, the smaller the nanoribbon width and hence higher the bandgap [15].

The GNRs are narrow strips of graphene nanosheet of finite width possessing structural similarity to that of unrolled carbon nanotube (CNT). GNRs possess mainly two types of structural configurations termed as armchair graphene nanoribbon (A-GNR) and zigzag graphene nanoribbon (Z-GNR). A-GNRs exhibit semiconducting properties whereas Z-GNRs are metallic in nature [15]. This behavior exclusively depends on how the graphene sheet cuts along with its plane (**Figure 1**). As shown in **Figure 3**, the bandgap increases with reducing nanoribbon width in an exponential manner [17–20]. Recently, it has been reported that applying a transverse magnetic field to the ribbon width induce the tunable bandgap i.e. tuning of the bandgap is possible by changing the magnitude of the applied

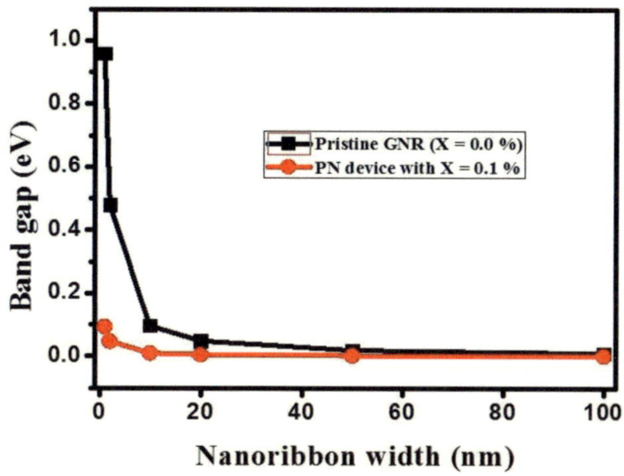

Figure 3.
The energy bandgap varies with nanoribbon width and doping concentration [17].

magnetic field [21]. The important aspect of a variable bandgap of a material is to develop efficient and flexible optoelectronic devices and sensors that work with the utmost accuracy. Considering this background, the subsequent sections discuss the photonic devices at the nanoscale in the present chapter.

2. Photonic devices at nanoscale

Photonic devices are the components that generate or detect the photonic flux that is developed and utilized for either electronic signal or light. The P-N device in the form of a light-emitting diode, photovoltaic cell, and the laser device is the most common type of photonic device. Traditionally these devices are designed and fabricated using Si or Ge due to which limited efficiency, broad bandwidth, high power consumption are few of the major limitations of current electronic devices. These limitations may be overcome by developing the replica of current electronic components at the nanoscale. The main theme of devices at the nanoscale is the smaller the dimensions, the lesser the power consumption, and the higher the efficiency. As mentioned above, though graphene possesses a zero energy bandgap in its nanosheet form, it exhibits a finite bandgap in its nanoribbons form which varies with the width of the nanoribbon. **Figure 3** shows that the bandgap of pristine graphene reduces exponentially with the increase in nanoribbon width.

It is possible to further enhance electronic transport properties of GNR with dopant adatoms [22–26] – that may help to fabricate graphene-based P-N nanodevices [27] as well as the surface acoustic wave sensors [28] at nanoscale. Moreover, it has also been proposed by several research groups that the doping of boron and nitrogen in graphene exhibits the possibility of engineering the graphene-based p-n junction at nanoscale as well as graphene aerogels for oxygen electro-catalysis [29, 30] wherein boron being trivalent and nitrogen being pentavalent impurities introduce the energy bandgap. **Figure 4** shows the boron and nitrogen-doped graphene nanoribbon based forward-biased p-n device. First-principles quantum transport calculations of electronic properties of boron and nitrogen-doped armchair GNR showed that the B-doped p-type GNR based device can exhibit high levels of performance, with high ON/OFF ratios and low subthreshold swing [31].

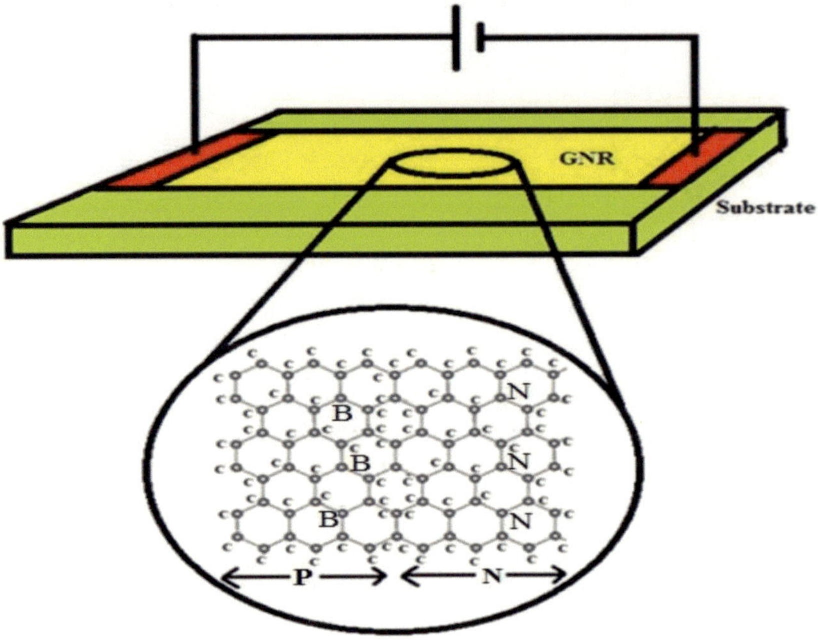

Figure 4.
Graphene-based P-N device [17].

2.1 Graphene-based P-N device and field effect transistor

The P-N device is one of the fundamental devices of an electronic circuit that controls the charge carrier (electron) current in the circuit manufactured from semiconducting materials such as silicon (Si) and germanium (Ge). It has a positive (p) region and negative (n) region created via doping semiconductor material by trivalent and pentavalent impurities respectively. Since AGNR exhibits semiconducting properties it is possible to design P-N device at the nanoscale using AGNR configuration. This type of P-N device will be having better electronic transport properties compared to traditional one because armchair graphene is not only a semiconductor but transparent and flexible also due to which it can be placed in nanoelectronic circuit. In addition to this, according to the recent article on graphene-based terahertz frequency detection, it is possible to design and fabricate graphene p-n junction based nano-antenna (bolometer) using the photo-thermoelectric effect wherein it is reported that with the dual gated dipolar antenna of the gap of 100 nm it is possible to concentrate the incident radiation for better photoresponse [32]. Graphene based field-effect transistors (GFET) are being investigated for more than a decade [33, 34].

There are several reasons behind this hunt such as limited electronic transport parameters of current electronics materials (Si, Ge) i.e. electron mobility and hence conductivity, poor heat dissipation rate of Si and Ge, and their tensile strength, failure of Moore's law, etc. In this context, graphene possesses superiority among all the materials known to researchers because graphene exhibits better electrical, mechanical, thermal, and optical properties in comparison with Si and Ge which are listed as follows: electrical conductivity of graphene is 10^7 S/m [35] whereas for Si it is 10^3 S/m, the electrical mobility of graphene is 10^5 cm^2/Vs [36] whereas for Si it is 10^3 cm^2/Vs, and 4×10^3 cm^2/Vs for Ge, Young's modulus of graphene is around 1.2 TPa [37] whereas that for both Si and Ge is 1 MPa, the thermal conductivity of graphene is around 5000 W/mK [37] whereas it is 1300 W/mK for Si and

580 W/mK for Ge. Considering these values of various parameters, it is obvious that graphene as a material is far better than the traditional semiconductors Si and Ge. In addition to this, graphene is a flexible transparent conducting thin film, unlike Si and Ge, due to which graphene can be used to develop flexible and transparent electronic devices that are base of wearable electronics [37].

The field effect transistor (FET) is one of the most important and fundamental electronic device that uses electric field to control the current and possesses three electrodes source, drain, and gate. A semiconductor channel is connecting source and drain and the third one i.e. gate controls the current. Implementation of graphene field effect transistors (GFET) in sensors has large number of benefits over the bulk FET made from Si. As the silicon is bulk semiconductor, the charge carriers at the channel interface have difficulty to penetrate into the device which limits response sensitivity of the device. On the other hand, as the graphene possesses two dimensional structure, the sensitive channel is itself the surface that ultimately improves the surface sensitivity. In addition to this, the carrier scattering rate through graphene is much lower than that in the case of bulk semiconductors. Therefore, the carrier energy loss also much lower than that for the bulk semiconductors. The fabrication of GFET is possible on Si/SiO_2 substrate with metal contacts via chemical vapor deposition (CVD) technique. G. Fiori et al. explored the possibility of tunable gap GFET considering bandgap opening by applying vertical electric field and using atomistic simulations based on the self-consistent solution of the Poisson and Schrödinger equations within the non-equilibrium Green's function formalism [38]. The chemical and biological sensors based on GFET were investigated to show their sensitivity towards detection of protein of different charge types [39]. Such sensors are having relatively higher sensitivities for biomolecules. On the other hand, GFET based high temperature sensor has been reported that works up to 600°C with utmost accuracy wherein researchers calculated the resistivity of the device using semi-classical transport Equations [40]. Graphene succeeded to implant itself in the broad field of organic light emitting diode (OLED) which is one of the important parts of optoelectronics [41–43]. Traditional OLEDs have their applications in screens of computer, mobile phones and cameras. In general, indium tin oxide (ITO) is used as transparent conductive thin film which is brittle and not flexible. In addition to this, indium may diffuse into active layers of OLEDs [44]. These limitations may be overcome using graphene instead of ITO because the work function of both the materials is same (4.5 eV).

2.2 Graphene-based photodetector and photovoltaic devices

Photodetectors are significant optoelectronic devices that detect the optical flux by converting the absorbed optical energy into the electronic current. They are part of remote control, televisions and DVD players. The spectrum responded by detectors is entirely depends on the bandgap of the material of detector. The traditional photodetectors consist of IV or III-IV semiconducting materials that are suffering from long-wavelength limits because these materials do not respond to the optical energy if its energy is less than the bandgap. Hence, the particular material becomes transparent for that radiation. As a solution of this problem, the implementation of graphene is the better option as graphene absorbs from ultraviolet to terahertz range [45, 46]. Since, the response time of a photodetector depends on the carrier mobility, graphene based photodetectors (GPDs) can be ultrafast because graphene exhibits very high carrier mobility. It is also possible to utilize photo-thermoelectric effect for efficient GPDs. In photo-thermoelectric effect, the photon energy converts into the heat followed by photocurrent generation. This is an important attribute to the fields of graphene based optoelectronics, photo-thermocouple devices and photovoltaic applications [47].

Graphene based touch screen is an emerging field as well because graphene is transparent and conducting too. This is the reason why graphene transparent conducting films (GTCFs) are promising layers for touch screens of electronic device displays. Graphene being mechanically strong, with high chemical durability, non-toxic, and cheap is one of the ideal materials for displays. Traditional displays consist of ITO which is costly, wear-resistant, brittle and has limited chemical durability. Graphene-based touch panel display can be grown by screen printing by the CVD technique [48]. Thus, GTCF may be an important part of future flexible and efficient touch screens.

For decades, it is known to us that photovoltaic (PV) cells convert light into electricity which is the main theme of traditional solar panels that are using silicon (Si) or germanium (Ge). The energy conversion efficiency of these materials is limited to around 25% [49]. Moreover, since Si and Ge are not flexible materials it limits the flexible solar cells or panels which are important components of futuristic wearable electronics. These limitations can be overcome by using graphene-based PV cells for this aspect. Graphene plays multiple roles in a photovoltaic cell i.e. photoactive material, transparent as well as conducting (TC) layer, charge transport layer, and catalyst. Among all these exciting and promising applications of graphene, the terahertz (THz) photonics based on graphene is a promising field of research as well which is capable to develop high-performance terahertz devices operated in the region between 300 GHz to 10 THz. at 300 K [50].

3. Graphene-based metamaterial

The modern form of artificial substance Metamaterials (MMs) has recently been examined for their electromagnetic properties that are missing in typical natural materials [51, 52]. The different results such as negative refractive index [53], perfect lensing [54], bolometer [55] etc. are discovered when utilizing these properties. From the other horizon, owing to its exceptional electrical, electronics and optical properties, such as strong thermal power, wide carrier mobility, and extremely young module [56], have acquired considerable attention in the domain of the thin and lightweight metamaterial research. Graphene is a 2-dimensional, radioactive medium that offers electrical and optical control across a large spectrum of frequencies, such as THz [57–59] and GHz [60]. The graphene's conductivity can be managed by various parameters such as temperature, duration, dispersion rate, and chemical potential [58]. Several devices are suggested for complete absorption [59, 61–63] polarization-insensitive [64–67], broad-angle [64–68], tunability [69–72]. For the Terahertz area and the Microwave, others are examined. **Figure 5** shows the example of the graphene-based squared shaped spiral metamaterial design for the polarization application. It is very much essential to discuss the refractive index parameters to identify the effect of metamaterial and negative refraction. The effective refractive index parameters can be identified by considering the transmittance and reflectance values of any two-port devices. The negative index behavior of any structure realized the overall effect of the metamaterial at a specific resonance frequency. The Schematic shown in **Figure 5** was numerically investigated to identify the reflection and transmission behavior of the THz wave. According to the results of transmittance and reflectance values, the effective refractive index was calculated for the band of 1 THz to 3 THz frequency region using the mathematical formula given in [73]. The negative refractive index response of the structure has been shown in **Figure 6**. **Figure 6** was derived for the different modes of excitation applied to the structure as shown in **Figure 5**. The effective refractive index of the normal and complimentary structure shows that the metamaterial effect was observed at the different resonance points for range 1 THz to 3 THz range. The proposed metamaterial

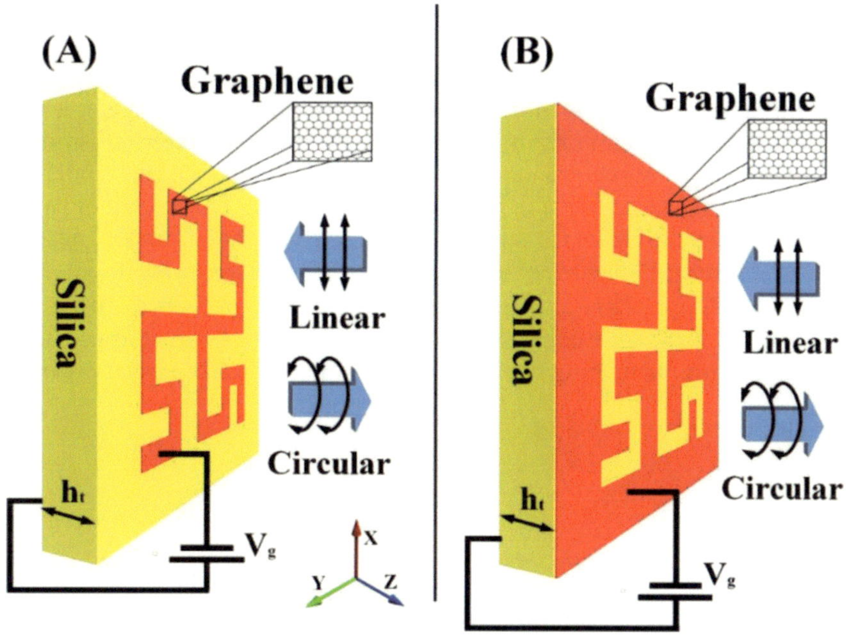

Figure 5.
Schematic of the squared spiral-shaped graphene metamaterial structure. (A) 3D view of the squared spiral-shaped graphene normal structure (SSSG – N), (B) squared spiral-shaped graphene complementary structure (SSSG – C) [73].

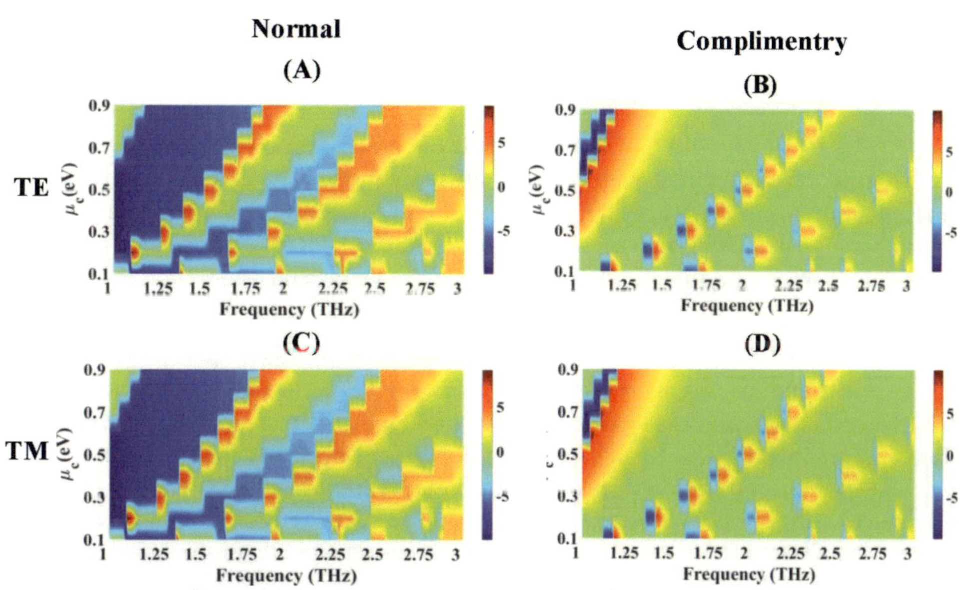

Figure 6.
The real part of the effective refractive index response of the proposed metamaterial structure for the TE and TM mode of the excitation over 1 THz to 3 THz frequency (structure shown in ***Figure 5****). The response is derived for the different chemical potential varied from 0.1 eV to 0.9 eV. (A) SSSG – N with TE excited mode, (B) SSSG – C with TE excited mode, (C) SSSG – N with TM excited mode and (D) SSSG – C with TM excited mode [74].*

behavior can help with different applications such as absorber, polarizer, superlens, etc. **Figure 6** also shows that for the different mode of the excitation (TE or TM) does not affect to the resonance behavior due to the squared spiral symmetric structure of the top graphene layer as shown in **Figure 5**.

4. Graphene surface conductivity model

Graphene-based photonics devices need to be analyzed by the specific mathematical characteristic before implementing it to the fabrication stage. It is important to identify the behavior of the graphene for the different external parameters such as temperature, frequency, external potential. Graphene can be modeled as one atom thick infinitesimally thin and two-sided surface. This model of the graphene can characterize by the surface conductivity model. Complex permittivity of the graphene sheet [75] is expressed by $\varepsilon(\omega)$ as expressed in Eq. 1, where the conductivity of the graphene σ_s expressed from Kubo formula as mentioned in Eqs. (2)–(4) [76]. Graphene conductivity is depending on the various parameters such as temperature, scattering rate, frequency and external chemical potential.

$$\varepsilon(\omega)=1+\frac{\sigma_s}{\varepsilon_0\omega\Delta} \tag{1}$$

$$\sigma_{intra}=\frac{-je^2k_BT}{\pi\hbar^2(\omega-j2\Gamma)}\left(\frac{\mu_c}{k_BT}+2\ln\left(e^{-\frac{\mu_c}{k_BT}}+1\right)\right) \tag{2}$$

$$\sigma_{inter}=\frac{-je^2}{4\pi\hbar}\ln\left(\frac{2|\mu_c|-(\omega-j2\Gamma)\hbar}{2|\mu_c|+(\omega-j2\Gamma)\hbar}\right) \tag{3}$$

$$\sigma_s=\sigma_{inter}+\sigma_{intra} \tag{4}$$

Here, ε_0 is the vacuum permittivity, σ_s is the monolayer conductivity, e is the fundamental electron charge value, ω is angular frequency, k_B is the Boltzmann's constant and $\hbar$ is the reduced Planck's constant. In this work, the chemical potential of graphene μ_c is varied between 0.1 eV to 0.6 eV, electron relaxation time τ^{-1} = 10-13 s, Γ is phenomenological scattering rate, graphene sheet thickness Δ = 0.34 nm and temperature T = 300 K. Carrier concentration of the whole graphene sheet will be controlled by field-effect, which can be indicated as: $n_s=\varepsilon_0\varepsilon_d V_g/et_d$ [77], where ε_0, ε_d = 2.25 and td = 2 μm are free space permittivity, permittivity of silica and thickness of silica layer respectively. Vg is gate voltage applied to graphene surface. Many research works have been used the graphene conductivity formula as it is easy to define in the computational studies with finite element method (FEM) of finite difference time domain method (FDTD). Simplified graphene conductivity module can be used also to identify the behavior of the entire structure for different physical parameters. It is also available several software packages that help the researchers to characterize the graphene material using mathematical modeling. RF module of the COMSOL Multiphysics provides the 2D and 3D analysis of the graphene-based devices for the photonics applications [77–79].

5. Conclusion

In summary, the present chapter discusses the current and future possibilities to incorporate wonder material graphene into the current photonic devices to enhance their performance in terms of efficiency, and sensitivity of sensors at the nanoscale replacing traditional semiconductors by graphene. Being a flexible, optically transparent, electrically as well as thermally conducting nanomaterial, graphene, in its nanosheet and nanoribbons forms, can be implanted into the current optoelectronic devices to overcome their existing limitations. However, the synthesis and insertion of graphene and its allotropes into the devices require state-of-the-art and is challenging too. We propose that the inclusion of graphene in traditional devices will take the current electronic devices to new horizons and will open up new frontiers of optoelectronic sensor technologies.

Author details

Ankur Pandya[1*], Vishal Sorathiya[2] and Sunil Lavadiya[2]

1 Department of Electronics and Communication, Nirma University, Ahmedabad, Gujarat, India

2 Department of Information and Communication Technology, Marwadi University, Rajkot, India

*Address all correspondence to: ankur.pandya@nirmauni.ac.in

IntechOpen

© 2020 The Author(s). Licensee IntechOpen. This chapter is distributed under the terms of the Creative Commons Attribution License (http://creativecommons.org/licenses/by/3.0), which permits unrestricted use, distribution, and reproduction in any medium, provided the original work is properly cited.

References

[1] Schwierz F. Graphene transistors. Nat. nanotechnol. 2010;5(7):487. DOI: org/10.1038/nnano.2010.89

[2] Geim A K. Graphene: status and prospects. Science. 2009;324:1530. DOI: 10.1126/science.1158877

[3] Zhu Y, James D K, Tour J M, New routes of graphene, graphene oxide and their related applications, Adv. Mater., 2012:24:4924.- DOI: 10.1002/adma.201202321

[4] D. Zhan, J. X. Yan, L. F. Lai, Z. H. Ni, L. Liu, Z. X. Shen, Engineering the electronic structure of graphene, Adv. Mater., 2012:24:4055.- doi:10.1002/adma.201200011

[5] C. Wang, D. Li, C.O. Too, G.G. Wallace, Electrochemical properties of graphene paper electrodes used in Lithium batteries, Chem. Mater.,2009:21: 2604, https://doi.org/10.1021/cm900764n

[6] F. Chen, J. L. Xia, D. K. Ferry, N. J. Tao, Dielectric screening enhanced performance in graphene FET, Nano Lett., 2009:9:2571,- https://doi.org/10.1021/nl900725u

[7] S. Abadal, I. Llatser, A. Mestres, H. Lee, Time-Domain Analysis of Graphene-Based Miniaturized Antennas for Ultra-Short-Range Impulse Radio Communications, IEEE Trans. Commun., 2015:63:1470- DOI: https://doi.org/10.1109/TCOMM.2015.2406691

[8] M. D. Stoller, S. J. Park, Y. W. Zhu, J. H. An, R. S. Ruoff, Graphene based ultracapacitors, Nano Lett., 2008:8:3498- https://doi.org/10.1021/nl802558y

[9] X. Wang, L. J. Zhi, N. Tsao, Z. Tomovic, J. L. Li and K. Mullen, Angew, Transparent carbon films as electrodes in organic solar cells, Chem., Int. Ed., 2008:47:2990.- https://doi.org/10.1021/nn400576v

[10] E. Yoo, T. Okata, T. Akita, M. Kohyama, J. Nakamura, I. Honma, Enhanced electrolytic activity at Pt subnanoclusters on graphene, Nano Lett., 2009:9: 2255.- https://doi.org/10.1021/nl900397t

[11] N. Ruecha, R. Rangkupan, N Rodthongkum, O. Chailapakul, Biosensors and Bioelectronics, 2014:52:13.- https://doi.org/10.1016/j.bios.2013.08.018

[12] Ye Lu, Mitchell B. Lerner, Zhengqing John Qi, Joseph J. Mitala Jr., Jong Hsien Lim, Bohdana M. Discher, and A.T. Charlie Johnson Jr. Graphene protein bioelectronic devices with wavelength-dependent photoresponse. Appl. Phys. Lett, 2012:100: 033110.- https://doi.org/10.1063/1.3678024

[13] Kumar V, Khandelwal G, Graphene-based Flexible and Stretchable Bioelectronics in Health Care Systems. J Anal Pharm Res, 2016:3(2):53- https://doi.org/10.1039/C9SM02127B

[14] Y Li, C H Park, Y W Son, M L Cohen, S G Louie, Quasiparticle energies and energy band gap in graphene nanoribbons, Phys. Rev. Lett. 2007:99: 186801- DOI: https://doi.org/10.1103/PhysRevLett.99.186801

[15] Y W Son, M L Cohen, S G Louie, Energy gaps in graphene nanoribbons, Phys Rev Lett., 2006:97,216803.- DOI :https://doi.org/10.1103/PhysRevLett.97.216803

[16] V Barone, O Hod, G Scuseria, Electronic structure and stability of semiconducting graphene nanoribbons, Nano Lett., 2006:6(12):2748-2754- https://doi.org/10.1021/nl0617033

[17] Pandya A, Jha P K, Electronic transport characteristics of a graphene nanoribbon based p–n Device, J. Electron Mater., 2019:48(9):5702- https://doi.org/10.1007/s11664-019-07388-z

[18] Sarma, S. D., Adam, S., Hwang, E. H., & Rossi, E. (2011). Electronic transport in two-dimensional graphene. Reviews of modern physics, 83(2), 407. https://doi.org/10.1103/RevModPhys.83.407

[19] Mukherjee S., Kaloni T.P., Electronic properties of boron and nitrogen doped graphene- a first principle study, J Nanopart Res, 2012:14:1059- https://doi.org/10.1007/s11051-012-1059-2

[20] S. Bruzzone, G. Fiori, Appl. Phys. Lett. 2011:99:222108- https://doi.org/10.1063/1.3665183

[21] Pandya A, Sangani K, Jha P K, Band gap determination of nanoribbons of Graphene, h- Boron nitride, phosphorene, silicene, stanene, and germanene, Journal of Physics D: Applied Physics, 2020:53(41)- https://doi.org/10.1088/1361-6463/ab9783

[22] L. Ci, L. Song, C. Jin, D. Jariwala, D. Wu, Y. Li, A. Srivastava, Z. F. Wang, K. Storr, L. Balicas, F. Liu, P. M. Ajayan, Atomic layers of hybridized boron nitride and graphene domains, Nat. Mater. 2010:9:430- https://doi.org/10.1038/nmat2711

[23] P. Rani, V. K. Jindal, Designing bandgap of graphene B and N dopants atoms, RSC Adv., 2013:3: 802 - https://doi.org/10.1039/C2RA22664B

[24] Q. Peng, A. R. Zamiri, W. Ji, S. De, (2012), https://arxiv.org/abs/1107.1448v2.- https://doi.org/10.1007/s00707-012-0714-0

[25] S. Lin, N. Tran, M. Lin, (2018), https://arxiv.org/abs/1801.07285.

[26] A Pandya, P K Jha, J. Electron. Mater. 2017:46:2340- https://doi.org/10.1007/s11664-016-5274-y

[27] B. Thakur, G. Zhou, J. Chang, H. Pu, B.Jin, X. Sui, X. Yuan, C. Yang, M. Magruder, J. Chen, Biosens Bioelectron, 2018:110:16- https://doi.org/10.1016/j.bios.2018.03.014

[28] I. Kuznetsova, V. Anisimkin, V. Kolesov, V. Kashin, V. Osipenko, S. Gubin, S. Tkachev, E. Verona, S. Sun, A. Kuznetsova, Sezawa Sensor Actuat. B- Chem., 2018:272:236- https://doi.org/10.1016/j.snb.2018.05.158

[29] L. Putri, B. Ng, W. Ong, H. Lee, W. Chang, S. Chai, J. Mater. Chem. A, 2018:7:3181 - https://doi.org/10.1039/C7TA09723A

[30] W. Chen, L Xu, Y. Tian, H. Li, K. Wang, Carbon, 2018:137:458- https://doi.org/10.1016/j.carbon.2018.05.061

[31] B. Huang, Physics Lett. A, 2011:375(4):845- https://doi.org/10.1016/j.physleta.2010.12.050

[32] Castilla, S., Terrés, B., Autore, M., Viti, L., Li, J., Nikitin, A. Y., Vitiello, M. S. Fast and sensitive terahertz detection using an antenna-integrated graphene pn junction,Nanoletters,2019:19(5):2765- https://doi.org/10.1021/acs.nanolett.8b04171

[33] Echtermeyer, T. J., Lemme, M. C., Baus, M., Szafranek, B. N., Geim, A.K., & Kurz, H. Non-volatile switching in graphene field-effect devices, IEEE Electron Device Letters, 2008:29(8):952-954-https://doi.org/10.1109/LED.2008.2001179

[34] Lemme M. C., Echtermeyer T. J., Baus, M., Kurz, H, A graphene field-effect device, IEEE Electron Device Letters, 2007:28(4):282-284- https://doi.org/10.1109/LED.2007.891668

[35] Marinho, B., Ghislandi, M., Tkalya, E., Koning, C. E., & de With, G. (2012). Electrical conductivity of compacts of graphene, multi-wall carbon nanotubes, carbon black, and graphite powder. Powder Technology, 221, 351-358. DOI- https://doi.org/10.1016/j.powtec.2012.01.024

[36] Novoselov, K. S., Fal, V. I., Colombo, L., Gellert, P. R., Schwab, M. G., & Kim, K. (2012). A roadmap for graphene. nature, 490(7419), 192-200.DOI- https://doi.org/10.1038/nature11458

[37] Cao, G. (2014). Atomistic studies of mechanical properties of graphene. Polymers, 6(9), 2404-2432 Doi- https://doi.org/10.3390/polym6092404

[38] Fiori, G., Iannaccone, G, On the possibility of tunable-gap bilayer graphene FET, IEEE Electron Device Letters, 2009:30(3):261-264- https://doi.org/10.1109/LED.2008.2010629

[39] Ohno Y., Maehashi, K., Matsumoto, K. Chemical and biological sensing applications based on graphene field-effect transistors, Biosensors and Bioelectronics,2010:26(4):1727-https://doi.org/10.1016/j.bios.2010.08.001

[40] Banadaki, Y. M., Mohsin, K. M., & Srivastava, A, A graphene field effect transistor for high temperature sensing applications. In Nanosensors, Biosensors, and Info-Tech Sensors and Systems (Vol. 9060, p. 90600F). International Society for Optics and Photonics, (2014).- https://doi.org/10.1117/12.2044611

[41] Son, D. I., Kwon, B. W., Park, D. H., Seo, W. S., Yi, Y., Angadi, B., Choi, W. K. Emissive ZnO–graphene quantum dots for white-light-emitting Diodes, Nat. nanotechnol., 2012:7(7): 465-471- https://doi.org/10.1038/nnano.2012.71

[42] Seo, J. T., Han, J., Lim, T., Lee, K. H., Hwang, J., Yang, H., & Ju, S. Fully transparent quantum dot light-emitting diode integrated with graphene anode and cathode. ACS nano, 2014:8(12):12476-12482- https://doi.org/10.1021/nn505316q

[43] Song, S. H., Jang, M. H., Chung, J., Jin, S. H., Kim, B. H., Hur, S. H., Jeon, S. Highly efficient light-emitting diode of graphene quantum dots fabricated from graphite intercalation compounds, Advanced Optical Materials, 2014:2(11): 1016- https://doi.org/10.1002/adom.201400184

[44] Bonaccorso, F., Sun, Z., Hasan, T. A., Ferrari, A. C. Graphene photonics and optoelectronics. Nature photonics,2010:4(9):611- https://doi.org/10.1038/nphoton.2010.186

[45] Wright, A. R., Cao, J. C. & Zhang, C. Enhanced optical conductivity of bilayer graphene nanoribbons in the terahertz regime. Phys. Rev. Lett.2009:103: 207401- https://doi.org/10.1103/PhysRevLett.103.207401

[46] Dawlaty, J. M. et al. Measurement of the optical absorption spectra of epitaxial graphene from terahertz to visible. Appl. Phys. Lett. 2008:93:31905- https://doi.org/10.1063/1.2990753.

[47] Xu, X., Gabor, N. M., Alden, J. S., van der Zande, A. M., & McEuen, P. L. Photo-thermoelectric effect at a graphene interface junction, Nano letters, 2010: 10(2): 562-566- https://doi.org/10.1021/nl903451y

[48] Bae, S., Kim, H. K., Sze, Y., Xu, X., Park, J. S., Zheng, Y. Kim, Y. J., 30 inch roll-based production of high-quality graphene films for flexible transparent electrodes. (2009) 10.1038/nnano.2010.132

[49] M. A. Green, K. Emery, K. Bcher, D. L. King, S. Igari, Progress in Photovoltaics 1999:7:321.

[50] Tredicucci, A., & Vitiello, M. S., Device concepts for graphene-based

terahertz photonics. IEEE Journal of Selected Topics in Quantum Electronics, 2013:20(1): 130-138.-DOI: 10.1109/JSTQE.2013.2271692

[51] Veselago VG (1968) THE ELECTRODYNAMICS OF SUBSTANCES WITH SIMULTANEOUSLY NEGATIVE VALUES OF $\ epsilon$ AND μ. Sov Phys Uspekhi 10:509-514. https://doi.org/10.1070/PU1968v010n04ABEH003699

[52] Smith DR, Padilla WJ, Vier DC, et al (2000) Composite medium with simultaneously negative permeability and permittivity. Phys Rev Lett 84:4184-4187. https://doi.org/10.1103/PhysRevLett.84.4184

[53] Shelby RA (2001) Experimental Verification of a Negative Index of Refraction. Science (80-) 292:77-79. https://doi.org/10.1126/science.1058847

[54] Singh R, Plum E, Zhang W, Zheludev NI (2010) Highly tunable optical activity in planar achiral terahertz metamaterials. Opt Express 18:13425. https://doi.org/10.1364/oe.18.013425

[55] Maier T, Brückl H (2009) Wavelength-tunable microbolometers with metamaterial absorbers. Opt Lett 34:3012. https://doi.org/10.1364/OL.34.003012

[56] Geim AK, Novoselov KS (2007) The rise of graphene. Nat Mater 6:183-91. https://doi.org/10.1038/nmat1849

[57] Huang F, Fu Y (2017) Theoretical T Circuit Modeling of Graphene-Based Metamaterial Broadband Absorber. Plasmonics 12:571-575. https://doi.org/10.1007/s11468-016-0299-x

[58] Hanson GW (2008) Dyadic green's functions for an anisotropic, non-local model of biased graphene. IEEE Trans Antennas Propag 56:747-757. https://doi.org/10.1109/TAP.2008.917005

[59] Dave V, Sorathiya V, Guo T, Patel SK (2018) Graphene based tunable broadband far-infrared absorber. Superlattices Microstruct 124:113-120. https://doi.org/10.1016/j.spmi.2018.10.013

[60] Ding F, Cui Y, Ge X, et al (2012) Ultra-broadband microwave metamaterial absorber. Appl Phys Lett 100:103506. https://doi.org/10.1063/1.3692178

[61] Thomas L, Sorathiya V, Patel SK, Guo T (2019) Graphene-based tunable near-infrared absorber. Microw Opt Technol Lett 61:1161-1165. https://doi.org/10.1002/mop.31712

[62] Mishra R, Sahu A, Panwar R (2019) Cascaded Graphene Frequency Selective Surface Integrated Tunable Broadband Terahertz Metamaterial Absorber. IEEE Photonics J 11:. https://doi.org/10.1109/JPHOT.2019.2900402

[63] Jia X, Wang X, Yuan C, et al (2016) Novel dynamic tuning of broadband visible metamaterial perfect absorber using graphene. J Appl Phys 120:033101. https://doi.org/10.1063/1.4956437

[64] Sorathiya V, Dave V (2020) Numerical study of a high negative refractive index based tunable metamaterial structure by graphene split ring resonator for far infrared frequency. Opt Commun 456:124581. https://doi.org/10.1016/j.optcom.2019.124581

[65] Sorathiya V, Patel SK, Katrodiya D (2019) Tunable graphene-silica hybrid metasurface for far-infrared frequency. Opt Mater (Amst) 91:155-170. https://doi.org/10.1016/j.optmat.2019.02.053

[66] Zeng L, Huang T, Liu GB, Zhang HF (2019) A tunable ultra-broadband linear-to-circular polarization converter containing the graphene. Opt Commun 436:7-13. https://doi.org/10.1016/j.optcom.2018.11.079

[67] Mattiucci N, Trimm R, D'Aguanno G, et al (2012) Tunable, narrow-band, all-metallic microwave absorber. Appl Phys Lett 101:141115. https://doi.org/10.1063/1.4757282

[68] Bai Y, Zhao L, Ju D, et al (2015) Wide-angle, polarization-independent and dual-band infrared perfect absorber based on L-shaped metamaterial. Opt Express 23:8670. https://doi.org/10.1364/OE.23.008670

[69] Guo Y, Zhang T, Yin WY, Wang XH (2015) Improved hybrid FDTD method for studying tunable graphene frequency-selective surfaces (GFSS) for THz-wave applications. IEEE Trans Terahertz Sci Technol 5:358-367. https://doi.org/10.1109/TTHZ.2015.2399105

[70] Wang DW, Zhao WS, Xie H, et al (2017) Tunable THz multiband frequency-selective surface based on hybrid metal-graphene structures. IEEE Trans Nanotechnol 16:1132-1137. https://doi.org/10.1109/TNANO.2017.2749269

[71] Yan R, Arezoomandan S, Sensale-Rodriguez B, Xing HG (2016) Exceptional Terahertz Wave Modulation in Graphene Enhanced by Frequency Selective Surfaces. ACS Photonics 3:315-323. https://doi.org/10.1021/acsphotonics.5b00639

[72] Li X, Lin L, Wu LS, et al (2017) A Bandpass Graphene Frequency Selective Surface with Tunable Polarization Rotation for THz Applications. IEEE Trans Antennas Propag 65:662-672. https://doi.org/10.1109/TAP.2016.2633163

[73] D.R. Smith, S. Schultz, P. Markoš, C.M. Soukoulis, determination of effective permittivity and permeability of metamaterials from reflection and transmission coefficients. Phys Rev B Condens Matter Mater Phys 65, 1-5 (2002). https ://doi.org/10.1103/Physrevb.65.19510 4

[74] Patel, S.K., Sorathiya, V., Lavadiya, S. et al. Numerical analysis of polarization-insensitive squared spiral-shaped graphene metasurface with negative refractive index. Appl. Phys. B 126, 80 (2020). https://doi.org/10.1007/s00340-020-07435-2

[75] Jiang, Q. Zhang, Q. Ma, S. Yan, F. Wu, X. He, Dynamically tunable electromagnetically induced reflection in terahertz complementary graphene metamaterials, Opt. Mater. Express. 5 (2015) 1962. doi:10.1364/ome.5.001962.

[76] G.W. Hanson, Dyadic Green's functions and guided surface waves for a surface conductivity model of graphene, J. Appl. Phys. 103 (2008). doi:10.1063/1.2891452.

[77] Y. Jiang, W.B. Lu, H.J. Xu, Z.G. Dong, T.J. Cui, A planar electromagnetic "black hole" based on graphene, Phys. Lett. Sect. A Gen. At. Solid State Phys. 376 (2012) 1468-1471. doi:10.1016/j.physleta.2012.03.018.

[78] Tutorial models for COMSOL Webinar "Simulating Graphene-Based Photonic and Optoelectronic Devices" [Internet]. 2020. Available from: https://www.comsol.co.in/community/exchange/361/ [Accessed: 2020-08-30]

[79] Alexander V. Kildishev, "Graphene Paves the Way for Next-Generation Plasmonics", [Internet]. 2020. https://www.comsol.com/story/graphene-paves-the-way-for-next-generation-plasmonics-53551 [Accessed: 2020-08-30]

Section 2

Photonic Devices

Chapter 3

Toward On-Demand Generation of Entangled Photon Pairs with a Quantum Dot

Arash Ahmadi, Andreas Fognini and Michael E. Reimer

Abstract

The generation of on-demand, optimally entangled photon pairs remains one of the most formidable challenges in the quantum optics and quantum information community. Despite the fact that recent developments in this area have opened new doors leading toward the realization of sources exhibiting either high brightness or near-unity entanglement fidelity, the challenges to achieve both together persist. Here, we will provide a historical review on the development of quantum dots (QDs) for entangled photon generation, with a focus on nanowire QDs, and address the latest research performed on nanowire QDs, including measuring entanglement fidelity, light-extraction efficiency, dephasing mechanisms, and the detrimental effects of detection systems on the measured values of entanglement fidelity. Additionally, we will discuss results recently observed pertaining to resonant excitation of a nanowire QD, revealing the potential of such sources to outperform spontaneous parametric down-conversion (SPDC) sources, providing a viable solution to the current challenges in quantum optics and quantum information.

Keywords: nanowire quantum dot, entanglement, dephasing, resonant two-photon excitation, fine-structure splitting

1. Introduction

Entangled photon pairs are one of the key elements for research and in emerging quantum applications with successful results in quantum foundations [1, 2], quantum communication [3–5], and quantum information [6–8]. Thus far, nonlinear crystals exhibiting spontaneous parametric down-conversion (SPDC) [9–11] have been the main source of generating entangled photon pairs for use in these areas. This type of source results in photon pairs that exhibit near-unity entanglement fidelity, high degrees of single-photon purity and indistinguishability in each emission mode, and high temporal correlation. Moreover, these sources perform at or near room temperature. However, there are fundamental limitations to such sources, which limit their performance and scalability for use in quantum photonics; an ideal source is imperative for optimal performance. One key feature of an ideal source of entangled photons is the ability to perform on-demand, i.e., source triggering and extraction of light must be possible with near-unity efficiency. SPDC sources follow a stochastic process and therefore generate entangled photon pairs at random. Moreover, the probability of multiphoton generation follows a Poisson distribution, and thus entanglement

IntechOpen

fidelity, single-photon purity, and photon indistinguishability [12] degrade when the pump power is increased [13]. As a result, these sources only operate at extremely low pair-production efficiencies, $\epsilon_p < 1\%$, per excitation pulse [14]. Hence, engineering and realizing an ideal source of entangled photons is necessary for the successful future of entangled photon pairs for use in quantum photonics, a future made brighter by semiconductor quantum dots (QDs).

Semiconductor quantum dots [15] are capable of generating pairs of entangled photons based on a process called the biexciton (XX)-exciton (X) cascade [16]; this cascade process is shown in **Figure 1**. The $|XX\rangle$ state is composed of two electron–hole ($e - h$) pairs in the QD's lowest energy level, i.e., s-shell. Each of these pairs possesses an angular momentum j_z with the superposition of $j_z = \pm 1$. From the $|XX\rangle$ state, there are two recombination pathways through the intermediate $|X\rangle$ state. Upon recombination, the $e - h$ pair will emit either a left, $|L\rangle$, or right, $|R\rangle$, circularly polarized photon corresponding to $j_z = +1$ or $j_z = -1$, respectively, and the QD final state at this step will be in the $|X\rangle$ state. This transition, $|XX\rangle \rightarrow |X\rangle$, is referred to as the neutral biexciton, XX, transition. Relaxation to the ground state occurs by the recombination of the remaining $e - h$ pair in the $|X\rangle$ state. The latter transition, i.e., $|X\rangle \rightarrow |G\rangle$, is referred to as the neutral exciton, X, transition. This recombination will emit a photon with a polarization perpendicular to that of the first photon (XX), i.e., $|L\rangle_{XX} \rightarrow |R\rangle_X$ and $|R\rangle_{XX} \rightarrow |L\rangle_X$. At the end of the process, the two emitted photons will be in the polarization entangled state [16]:

$$|\Psi\rangle = \frac{1}{\sqrt{2}}(|RL\rangle + |LR\rangle). \tag{1}$$

Over the past three decades, QDs have been extensively studied with recent advancements, as compared to other solid state quantum emitters [18–21], and have produced sources which exhibit features closest to an ideal photon source [22]. The first generation of QDs was self-assembled [23–25], which resulted in QDs with various sizes and imperfect symmetry due to the random nature of the formation process [13]. Moreover, since the bulk semiconductor material possessed a high

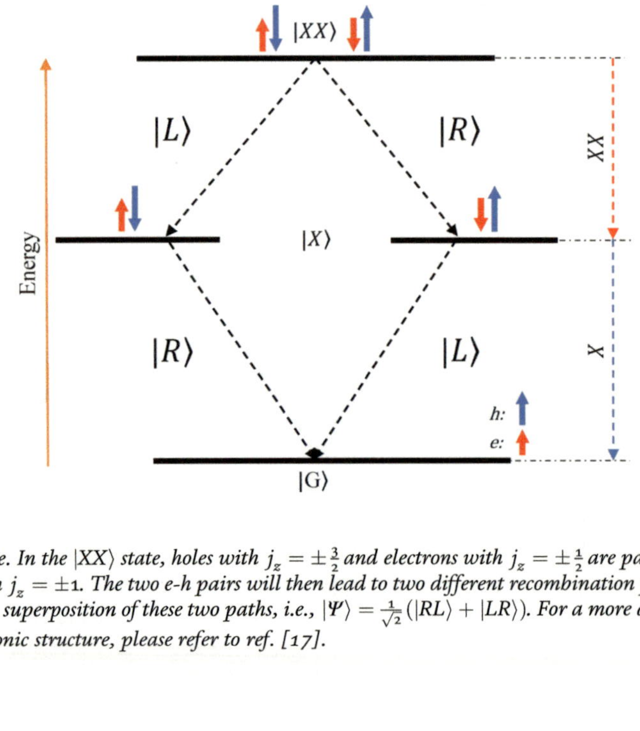

Figure 1.
The XX-X cascade. In the $|XX\rangle$ state, holes with $j_z = \pm\frac{3}{2}$ and electrons with $j_z = \pm\frac{1}{2}$ are paired, resulting in exciton states with $j_z = \pm 1$. The two e-h pairs will then lead to two different recombination pathways, with the final state being a superposition of these two paths, i.e., $|\Psi\rangle = \frac{1}{\sqrt{2}}(|RL\rangle + |LR\rangle)$. For a more detailed description of the QDs' electronic structure, please refer to ref. [17].

refractive index, these self-assembled QDs typically suffered from isotropic emission and total internal reflection at the semiconductor-air interface and thus exhibited a low light-extraction efficiency of $\sim 1\%$ [26].

Recent developments in micro- and nanoscale crystal growth and fabrication have resulted in structures which have improved the performance of QDs considerably. Enhancement of the spontaneous emission of QDs was first achieved by coupling an ensemble of QDs [27], and later a single QD, to a micro-cavity [28]. More recently, the coupling of QDs to micro-pillar cavities has achieved light-extraction efficiencies as high as 80% [29]. Also, such structures allow for proper control of the charge noise around the QD and thus the suppression of detrimental dephasing processes from the moving charge carriers. Excitingly, as a result, photons with $> 99\%$ indistinguishability and single-photon purity have been reported [30].

However, such performance comes at a price. Due to Coulomb interactions [17], XX and X emission lines are separated in energy by an amount referred to as the XX binding energy, Δ_{XX-X}. Within a typically used cavity with a quality factor of $Q \sim 10,000$, which is tuned to photons with the wavelength $\lambda \sim 1\mu$m, the cavity has a bandwidth of $\sim 10\mu$eV. This small bandwidth is far less than a typical XX binding energy of $\Delta_{XX-X} \sim 1$meV [31]. Therefore, either the X or the XX transition lines can be coupled to the fundamental mode of the cavity, but not both as needed for a high-efficiency entangled photon source. Furthermore, to avoid suboptimal entanglement, coupling to the fundamental mode of the cavity should be polarization-independent. Otherwise, one decay path in the cascade process would gain a stronger weight [32]. In an attempt to overcome these challenges, three studies have made considerable gains. Dousse et al. [32] fabricated a micropillar cavity molecule and successfully tuned both XX and X transitions to separate cavity modes in the molecule. Impressively, the fabricated device was shown to improve the pair-extraction efficiency (ϵ_e) by three orders of magnitude, $\epsilon_e = 12\%$, as compared to a bare self-assembled QD; yet despite this, the source still generates poorly entangled photon pairs with a measured fidelity of $\mathcal{F} = 67\%$. Other research by Chen et al. [26] showed the successful fabrication of a broadband dielectric antenna, which enables a pair-extraction efficiency of 37.2%, low multiphoton emission $g^{(2)}(0) \approx 0.002$, and entanglement fidelity, $\mathcal{F} = 90\%$; however, the emission profile deviates from a Gaussian, and photon indistinguishability of such sources is yet to be measured. Additionally, Wang et al. [33] have engineered a circular Bragg grating bull's-eye cavity around a single QD showing a pair-extraction efficiency, $\epsilon_e = 36.6\%$, and an entanglement fidelity, $\mathcal{F} = 90\%$. Despite these impressive gains, still none exhibit the promise of quantum dots in realizing an ideal entangled photon source, showing near-unity entanglement fidelity and pair-extraction efficiency.

Another important feature of QDs affecting the measured entanglement is the fine-structure splitting (FSS) of the $|X\rangle$ state, which is caused by the exchange interaction of the electron and hole in the $e-h$ pair, together with geometrical asymmetries of the QD [34, 35]. As a result, in the $XX-X$ cascade, there is a precession between the two recombination pathways in the $|R\rangle/|L\rangle$ basis (**Figure 2(a)**), and a non-degeneracy in energy of the two pathways in the $|H\rangle/|V\rangle$ basis will appear (**Figure 2(b)**). Therefore, the two-photon quantum state described by Eq. (1) will change into:

$$\begin{aligned} |\tilde{\Psi}\rangle &= \frac{1}{\sqrt{2}}\left(|HH\rangle + e^{i\frac{\delta}{2\hbar}t}|VV\rangle\right) \\ &= \cos\left(\frac{\delta}{2\hbar}t\right)\Psi + i\sin\left(\frac{\delta}{2\hbar}t\right)|\Phi\rangle, \end{aligned} \tag{2}$$

where $|\Psi\rangle$ is the state described by Eq. (1), $|\Phi\rangle = \frac{1}{\sqrt{2}}(|RR\rangle + |LL\rangle)$, and $\delta = \text{FSS}$.

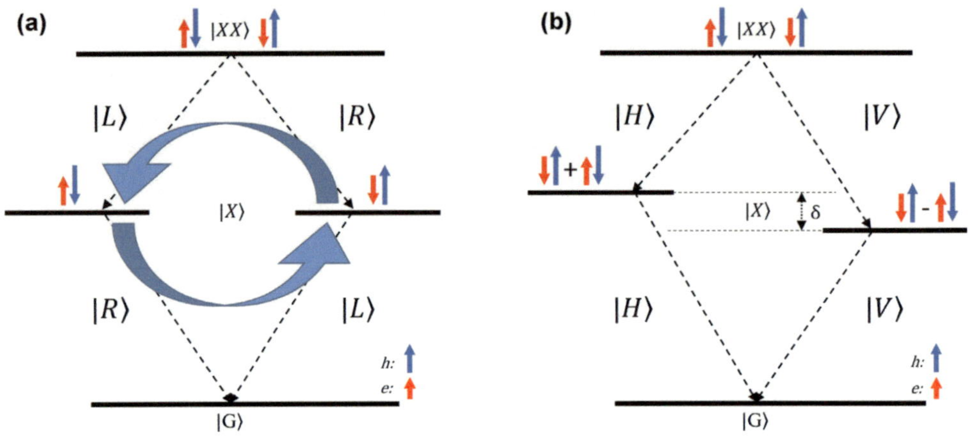

Figure 2.
XX-X cascade in the presence of FSS. (a) $|R\rangle$ and $|L\rangle$ basis will be mixed and a precession between the two pathways will be observed. (b) In the $|H\rangle/|V\rangle$ basis, the transition energies will be split by FSS $= \delta$.

Due to the random nature of the growth process, self-assembled QDs have long suffered from large base asymmetries, which resulted in FSS values larger than the X emission linewidth. This feature will lead to the introduction of a which-path information in the $XX - X$ cascade that will degrade the entanglement between the two emitted photons. For this reason, the early measurements on entanglement in QDs [36, 37] only led to detection of classical correlations; nonclassical correlations were only observed by improving the growth techniques and choosing QDs with FSS ≈ 0 [38, 39]. In the recent past, several techniques have been proposed and demonstrated in order to erase the FSS of QDs using electric fields [40–42], strain [43], and an optical approach not requiring nanofabrication [44]. However, as we show further along in the chapter, in order to reveal the effect of FSS on entanglement fidelity of the emitted photon pair, a delicate understanding of the detection system is also required. A recent study by Fognini et al. [45] shows that it is possible to measure near-unity entanglement fidelity even in the presence of finite FSS.

To reveal the true potential of QDs, proper excitation schemes are needed in addition to engineering sophisticated photonic structures. Until recently, off-resonant excitation had been widely used to generate entangled and single photons from QDs in photonic structures. This scheme excites charge carriers to energy levels above the bandgap of the host semiconductor, and relaxation of the resulted $e - h$ pairs to the QD's s-shell, mediated by interactions with phonons, leads to the emission of entangled photons. Admittedly, implementing this scheme is relatively straightforward, as the large difference in the frequencies of the excitation laser and the emitted photons allow for simple filtration of the reflected laser light. The excess of charge carriers and their interaction with phonons will lead to detrimental effects such as inhomogeneous broadening of emission lines [46], multiphoton emission caused by re-excitation processes [47], increased jitter in emission time [48], and dephasing [45].

Direct population of $|XX\rangle$ is forbidden due to optical selection rules. However, observation of resonant two-photon absorption in photoluminescence excitation spectroscopy of QDs [49] has recently led to development of a resonant two-photon excitation (TPE) scheme [50–52], which allows for coherent population of the $|XX\rangle$ state. In order to perform this scheme (**Figure 3**), a linearly polarized excitation pulse is tuned to a virtual state with an energy halfway between that of the ground state, $E_{|G\rangle}$, and biexciton state, $E_{|XX\rangle}$. This virtual state can also be thought of as a transition level between that of the neutral exciton and that of the neutral biexciton transitions. In other words, $|XX\rangle$ is coherently populated by a laser pulse, which is

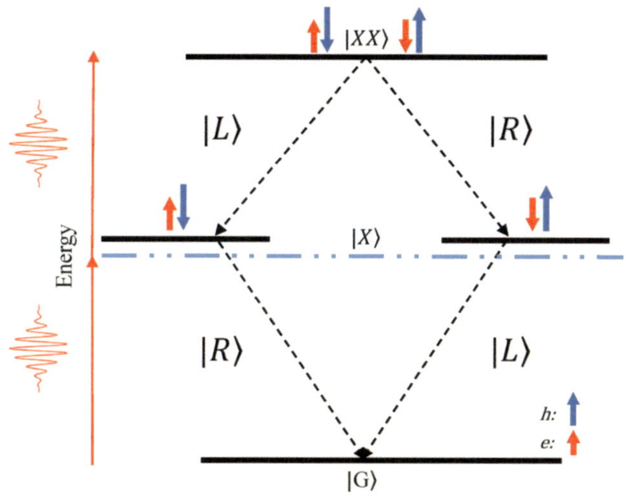

Figure 3.
Schematics of resonant TPE. A linearly polarized pulse is tuned to a virtual state halfway between X and XX transitions (the dashed blue line); and the $|XX\rangle$ is coherently populated via a two-photon absorption process.

resonant to neither X nor XX transitions. This method can lead to near-unity population of the $|XX\rangle$ state [53] and an extreme suppression of multiphoton emission of X and XX transitions [54].

In this review, we focus on attempts to improve the performance of entangled photon generation in by embedding them in photonic nanowires, as well as the effects of different excitation schemes in the performance of such sources. Additionally, we will also cover the improvements achieved in photon extraction efficiency, reduction of the dephasing processes, suppression of multiphoton emission, and enhancing entanglement fidelity of nanowire QD based entangled photon sources.

2. Nanowire QDs

Embedding QDs in tapered nanowires was initially developed by using top-down approaches via reactive-ion etching [55, 56]. Such photonic structures allow for coupling of the QD emission to the waveguide's fundamental mode in a broad range of wavelengths, $\Delta\lambda \approx 70$nm. Claudon et al. [55] managed to achieve a light-extraction efficiency of $\epsilon \approx 72\%$; however, top-down approaches are not flawless. Defects are left at the surface of the nanowire due to etching of the substrate using reactive ions, and additionally there is limited control in the positioning of the QDs at the symmetry axis of the nanowire. It is important to note that these flaws lead to suboptimal quality in the ultimate brightness of the source. As an alternative growth approach in attempt to overcome these issues, pure wurtzite InP nanowires were grown with a bottom-up growth approach and the quantum dot was placed on the nanowire axis to ensure good ($\sim$95%) coupling between the quantum dot and fundamental mode of the nanowire waveguide.

2.1 Bottom-up grown tapered wurtzite nanowire QDs

A novel bottom-up approach to growing tapered nanowires was used in the work by Reimer et al. [57]. This innovative approach allowed, for the first time, the positioning of a QD on the symmetry axis of the nanowire and at a desired height with a precision of $\sim$100 nm (**Figure 4**). In this method, the growth of the nanowire core, InP, is initiated by a gold particle which defines the core of the nanowire and ultimately the size of the QD, $D \approx 20 - 30$nm. After reaching the desired height, arsine is introduced to the growth chamber, and the QD, InAsP, is grown. Then, by changing the growth conditions, the nanowire is grown radially in order to create a shell $D \approx 220$nm, which facilitates the waveguide effect. In the last phase of growth, the conditions are changed once again to achieve an ideal tapering at the nanowire tip with an angle of $\theta < 2^{\circ}$, which results in minimal internal reflection for the emitted photons leaving the nanowire.

2.2 Optical properties

In terms of brightness, a value of $\epsilon = 43(4)\%$ for success probability of single-photon extraction at the first lens has been reported [46]. However, theoretically these sources allow for single-photon extraction efficiencies up to $\epsilon = 97\%$, which can be achieved by perfect tapering of the nanowire, $\theta = 1^{\circ}$, placing the nanowires on top of a flawless mirror and placing the QD at the correct height in order to create perfect constructive interference [57]. In terms of multiphoton emission (**Box 1**), second-order correlation, $g^{(2)}(0)$, measurements yield values $< 1\%$ [45, 46], showing a true single-photon emitter. The emitted photons exhibit extremely narrow linewidths, $\delta\omega < 1$GHz, with coherence lengths exceeding 1ns. Furthermore, a high level of visibility in a Hong-Ou-Mandel (HOM) measurement, $V = 85\%$, has been observed, indicating highly indistinguishable photons [46]. Moreover, there is a close to perfect overlap, $98.8\% \pm 0.1\%$, between the far-field emission profile of these nanowires QDs and a Gaussian emission profile of a

1. Hanbury Brown and Twiss (HBT) setup

In order to quantify the multiphoton emission of a source, the second-order correlation function is measured based on a setup first introduced by Hanbury Brown and Twiss [59] (**Figure a**). In this method, the light emitted from the source is sent to a beam splitter and then detected by two single-photon detectors D1 and D2. By correlating the intensities recorded by the two detectors in different time bins, one can gain information about the emission pattern of the source. Considering the particle nature of photons, if the source emits one and only one photon in each emission mode upon excitation, there will be no simultaneous detection on the two detectors; in other words, there will not be any correlation at zero time delay:

$$g^{(2)}(0) = \frac{\langle n_1(t) n_2(t) \rangle}{\langle n_1(t) \rangle \langle n_2(t) \rangle} = 0, \tag{3}$$

with $n_i(t)$ being the number of photons detected by detector $i\,(i = 1, 2)$ at time t.

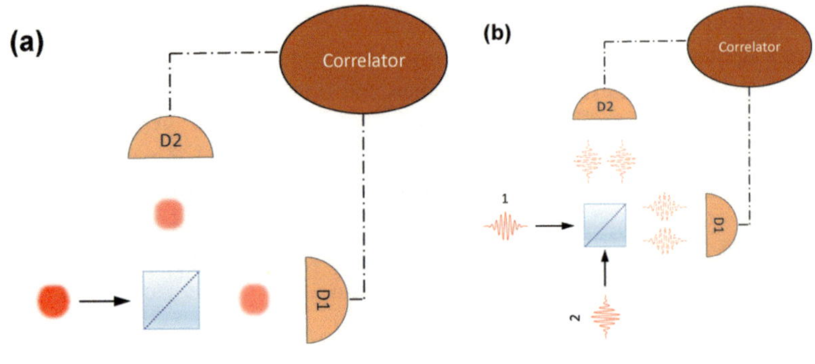

2. Hong-Ou-Mandel setup

In addition to single-photon emission, for an ideal entangled photon source, the emitted photons in each mode should exhibit perfect indistinguishability. For measuring this feature, the Hong-Ou-Mandel setup is used. Using a setup similar to that the HBT (**Figure b**) and considering the wave nature of the photons, a HOM measurement enables one to test the degree of indistinguishability of the successive photons. In this scenario, two successive photons are brought together at the beam splitter for interference. Now, at the beam splitter, four different possibilities exist (**Figure c**); photon 1 may be reflected and photon 2 transmitted (case 1), photon 1 may transmit and photon 2 be reflected (case 2), both may transmit (case 3), and, lastly, both may be reflected (case 4). With reflection from the two sides of the beam splitter differing in a π phase shift, the third and the fourth cases are physically identical except for a general phase. Therefore, in the interference of the two photons, these two cases will cancel out, leaving only the options with the two photons going to either D1 or D2. Building a histogram out of the correlations of the two detectors, similar to the second-order correlation measurement, exhibits zero coincidence counts at zero time delay for the case of a source generating perfectly indistinguishable photons in each mode.

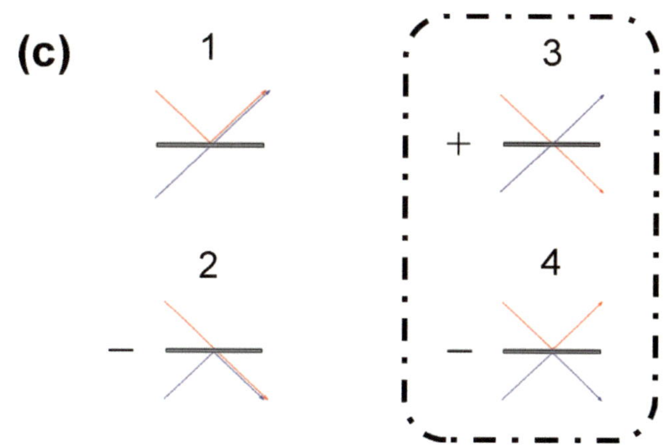

Box 1.
Measuring multiphoton emission and photon indistinguishability of entangled photon sources.

single-mode fiber; in practice, these sources have resulted in a coupling efficiency over 93% into a single-mode fiber [58]. Impressively, this feature then allows for possibilities in long-distance fiber-based quantum communication with high efficiency through low-loss communication channels.

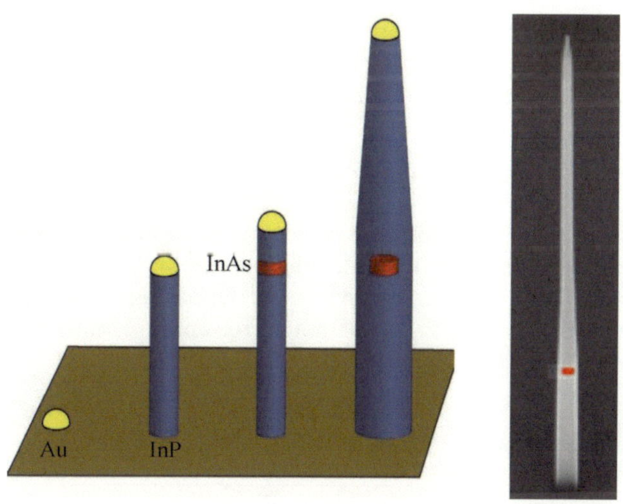

Figure 4.
Schematic of the bottom-up nanowire growth process and SEM image of a tapered nanowire (right). The growth process is initiated by a gold particle, which defines the dimensions of the QDf. After the quantum dot is grown the waveguide shell and the tapered tip are fabricated around the QD by controlling the growth parameters. This growth process ensures that the QD is placed on-axis of the tapered nanowire waveguide for efficient light extraction.

2.3 Entanglement measurements

Following the method introduced by James et al. [60], the first results in measuring the degree of entanglement in bottom-up grown nanowire QDs were reported in 2014 by Versteegh et al. [61]. In this work, using an above-bandgap excitation scheme, the fidelity of the emitted $XX - X$ pairs to a maximally entangled state was found to be $F = 0.859\,(\pm 0.006)$, with a concurrence equal to $C = 0.80\,(\pm 0.02)$, under strong post-selection conditions. The fidelity is reduced to 0.765 ± 0.002 with inclusion of 100% of the emitted photons. By changing the excitation conditions to excite the QD at the wurtzite InP nanowire bandgap, Jöns et al. [62] enhanced the fidelity of the same source as used by Versteegh et al. [61] to $F = 0.817 \pm 0.002$ by including all of the collected photon pairs. This strong degree of entanglement allowed Jöns et al. [62] to perform a Bell type inequality violation test, specifically the Clauser-Horne-Shimony-Holt (CHSH) measurement [63]. The CHSH measurement yielded a violation of Bell's inequality by 25 standard deviations, clearly showing the promising features of bottom-up nanowire QDs for secure quantum communication purposes. The experimental setup can be seen in **Figure 5**. Initially, a pair of $\lambda/2$ and $\lambda/4$ waveplates corrects for the birefringence observed in the nanowire, causing the entangled state to rotate to an elliptical state instead of the expected $|\Psi\rangle = \frac{1}{\sqrt{2}}(|HH\rangle + |VV\rangle)$ [61]. The stream of emitted photons are separated by a 50/50 beam splitter and sent to two separate detectors tuned to the specific wavelengths of the X and XX. In order not to be affected by the phase introduced when photons hit the reflecting surface of the beam splitter, the $\lambda/2$ and $\lambda/4$ waveplates used for projection measurements are aligned along the transmission path [62].

It is important to note that neither of the above-mentioned works addresses the ultimate entanglement fidelity achievable for nanowire QDs. In addition to the projection measurements, a more in-depth analysis is needed in order to reveal the underlying physical mechanisms such as dephasing due to nuclear spins and charge carriers through spin-flip processes. Moreover, the effect of FSS on the measured value of entanglement fidelity deserves more care, since, even though lifting of the degeneracy between the two decay paths in the $XX - X$ cascade can be interpreted as an introduction of a which-path information, the effect is purely unitary, and the precession shall not destroy the entanglement alone. Here, the detection apparatus will play a major role regarding the effect of FSS on the entanglement fidelity.

2.3.1 Dephasing-free entanglement in nanowire QDs

In an attempt to shed light on these finer aspects of generation of entangled photons in nanowire QDs, Fognini et al. [45] studied an InAsP QD embedded in an InP photonic nanowire, revealing the effects of dephasing, FSS, and imperfections of the detection system on the values achieved for entanglement fidelity.

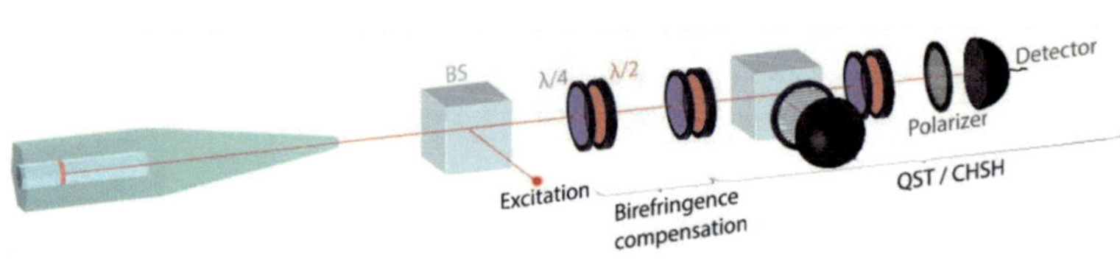

Figure 5.
Two-photon quantum state tomography setup. The setup consists of two pairs of $\lambda/4$–$\lambda/2$ wave plate sets, which combined with a pair of polarizors perform the projection measurements. A combination of $\lambda/2$ and $\lambda/4$ wave plates is used to compensate for the birefringence, if it is present in the nanowire (the image is taken from Jöns et al. [62]).

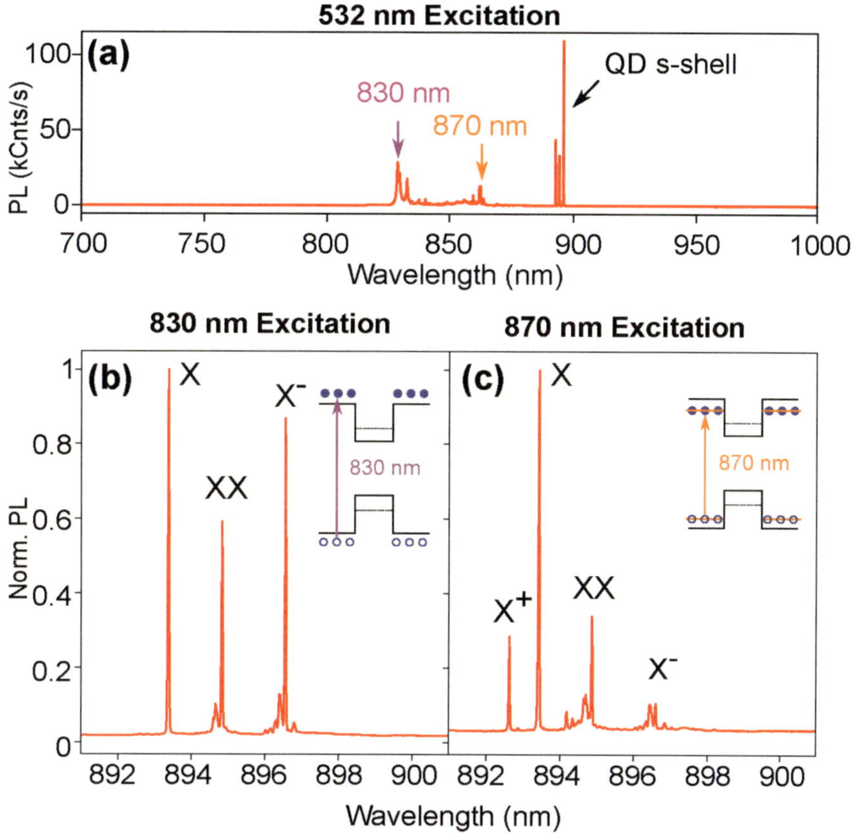

Figure 6.
QD emission spectra. (a) Emission spectrum by excitation via a green laser. Excitations at two different energy levels, wurtzite InP bandgap at 830 nm and donor/acceptor levels at ≈ 870 nm, were used for performing entanglement measurements; (b) the emission spectrum for 830 nm excitation exhibiting the exciton (X), biexciton (XX), and negatively charged exciton (X^-) lines. (c) Excitation at to 870 nm leads to an appearance of a positively charged exciton (X^+) and suppression of X^-. The spectra in (b) and (c) were taken at the saturation power of X.

The emission spectrum of the source is provided in **Figure 6**. Upon excitation of the sample using a green laser at $\lambda = 520$nm (**Figure 6a**), three sets of peaks can be observed: the wurtzite InP bandgap at $\lambda = 830$nm, levels attributed to the donors and acceptors which are formed due to the presence of impurities such as beryllium in the growth chamber, at $\lambda \approx 870$nm, and the *s*-shell of the QD at $\lambda \approx 894$nm. The $XX - X$ cascade can be generated by exciting the sample either at the InP bandgap or the donor/acceptor levels. The charge environment around the QD is different in the two cases. Whereas excitation at the InP bandgap (**Figure 6b**) leads to appearance of a negatively charged exciton (X^-), exciting the QD at the donor/acceptor level will lead to emission of positively charged excitons (X^+) and suppression of the X^- emission line. Moreover, using the donor/acceptor levels to excite the quantum dot, this excitation scheme will fill the charge traps around the QD. As a consequence, the charge mobility will be significantly reduced; a phenomenon which we will show to be extremely effective in suppressing the dephasing caused by the surrounding charge carriers.

Following a similar setup to the one used by Jöns et al. [62] (**Figure 5**), Fognini et al. [45] conducted two-photon quantum state tomography on the $XX - X$ cascade in time intervals of 100 ps during the decay time of the exciton, which allowed for the construction of the density matrix of the photon pair, and gave the opportunity to observe the evolution of the two photon quantum state. The result of these

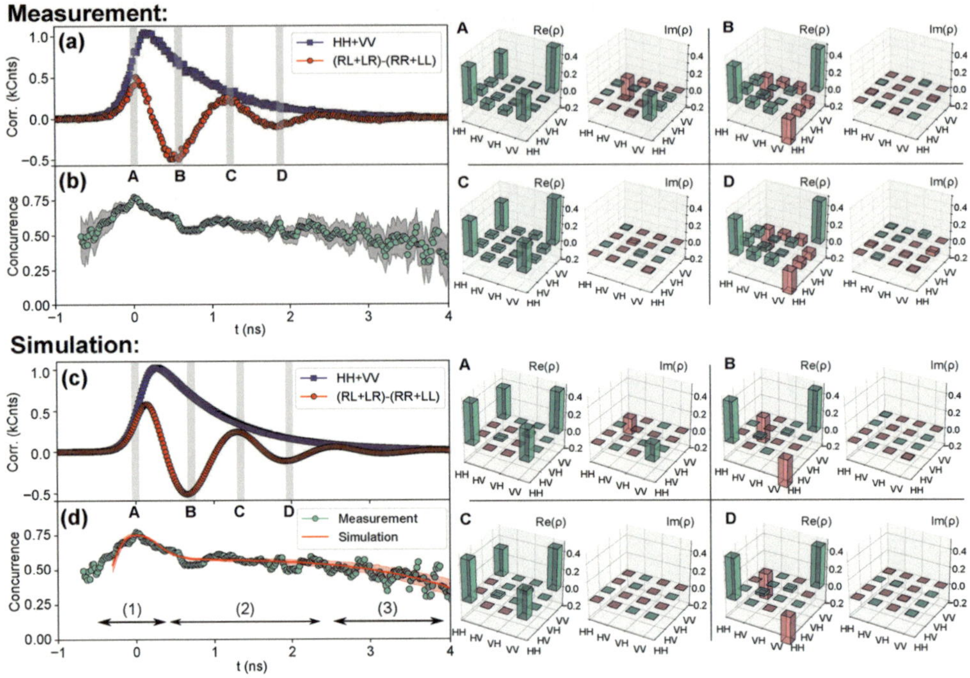

Figure 7.
Dephasing-free entanglement (a) showing the correlation measurements $HH + VV$ and $(RL + LR) - (RR + LL)$. The former does not show any oscillations as $|H\rangle$ and $|V\rangle$ are the eigenstates of the Hamiltonian, whereas the latter reveals the precession of the state between $|\Psi\rangle$ and $|\Phi\rangle$, according to Eq. (2). The shaded gray bars indicate instances with the highest concurrence, (A), and instances with the lowest imaginary part in the density matrix (B-D). (b) The concurrence extracted from the correlation measurements at each instant of time, for time windows of $\Delta t = 100$ ps, along the decay time of the exciton. (c) Result of the correlations obtained from the theoretical model (Eq. (4)) with the gray shaded bars indicating similar instances as for (a). (d) Comparison of the measured values of the concurrence with that of the theoretical model, revealing the dephasing free nature of the $XX - X$ cascade.

measurements with excitation at the donor/acceptor levels can be observed in **Figure 7a, b**. In **Figure 7a**, the results of the correlation measurements in the H/V basis and R/L basis are presented. By setting the detector tuned to the XX emission line as the "start," and the other detector tuned to X line as the "stop" in the correlation measurements, variations can be found within the coincidence counts in different bases during the exciton decay. Plotting the correlation counts $(RL + LR) - (RR + LL)$ vs. decay time reveals the precession of the two-photon quantum state between the two entangled states $|\Psi\rangle$ and $|\Phi\rangle$, according to Eq. (2), with a frequency proportional to FSS. From a fit to the measured oscillation of the two-photon quantum state, we calculate the FSS to be $\delta = 795.52 \pm 0.35$ MHz. From the $HH + VV$ correlation, a fit to the data yields an X lifetime of $\tau_X = 847 \pm 6$ ps. **Figure 7b** shows the results of calculating the concurrence [64], C, of the two-photon quantum state, with $C = 0$ indicating no entanglement and $C = 1$ showing a maximally entangled state, for time windows of 100 ps along the exciton decay time. The gray bars indicate the time instances when the calculated concurrence is the highest, A, and when the imaginary part of the density matrix is zero, B-D. The respective density matrices of each instance is given in the subplots on the top right of **Figure 7**. The concurrence reaches a value of $C = 0.77 \pm 0.02$ at its peak, corresponding to a fidelity of $\mathcal{F} = 0.88$, with a count-weighted average of $\overline{C} = 0.62 \pm 0.03$.

Despite the fact that the value for concurrence does not reach near unity and that after a peak around $t = 0$, it suffers a significant reduction, the observed behavior does not indicate the presence of a dephasing mechanism during the decay time of the exciton. Fognini et al. [45] further investigated the change of the concurrence value during the exciton's emission time as compared to the measured values with a model that included the parameters of the $XX - X$ cascade, $FSS, g^{(2)}(0)$ of X and XX, τ_X, etc., as well as the features of the detection system, including the detectors' timing resolution and dark counts, but did not include any term for dephasing.

Starting with the state described by Eq. (2), the expected values for 36 possible projection correlations, N_{ij} $(i, j \in \{H, V, D, A, R, L\})$ with the letters indicating the photon polarization along horizontal, vertical, diagonal, antidiagonal, right, and left, respectively, at time t and during a time interval Δt can be written as:

$$N_{ij}(t) = N_0\left(|\langle ij|\Psi(t,\delta)\rangle|^2 n(t,\tau_X)\right) * g(t)\Delta t \tag{4}$$

where N_0 is the total number of photon pairs collected, δ is the value of FSS, τ_X is the lifetime of the exciton state, $n(t,\tau_X) = 1/\tau_X\left(e^{-t/\tau_X}\right)$ describes the emission probability of an exciton following an exponential decay, $*$ is the convolution operator, and $g(t)$ denotes the detectors' timing resolution function.

To construct the density matrix of the two-photon quantum state, Eq. (4) gives the correlations in all 36 bases with the effect of the detectors' timing resolution function included. However, two additional factors should be included, $g^{(2)}(0)$ of X and XX, and also the detectors' dark counts. The dark counts will result in detection of false correlations that are evenly distributed in time, which has to be added to the raw correlations obtained by Eq. (4). On the other hand, the system studied by Fognini et al. [45] exhibits values of $g_X^{(2)}(0) = 0.003 \pm 0.003$ for X and $g_{XX}^{(2)}(0) = 0.10 \pm 0.01$ for XX by inclusion of the counts in a range of $\Delta t = 100$ ps in the proximity of $t = 0$. A non-zero value of $g^{(2)}(0)$ for either XX or X results in the addition of uncorrelated photons in the tomography measurement and thus a reduction in the measured entanglement fidelity. Now, by only considering $g_{XX}^{(2)}(0)$, since $g_X^{(2)}(0)$ is negligible in comparison, one finds that uncorrelated photons are being detected in $g_{XX}^{(2)}(0)$ fraction of the times, which has to be taken into consideration. In other words, the density matrix constructed by considering the correlations in different bases, described in Eq. (4), and addition of the effect of the dark counts describe the behavior of the system only in $\left(1 - g_{XX}^{(2)}(0)\right)$ fraction of the times. Therefore, the actual density matrix is expected to be:

$$\rho_{sim}(t) = \left(1 - g_{XX}^{(2)}(0)\right)\rho_{raw}(t) + g_{XX}^{(2)}(0)\frac{\mathbb{I}}{4} \tag{5}$$

where $\rho_{sim}(t)$ is the density matrix of the two-photon quantum state at a partic ular time t based on the simulation and after considering all of the factors; $\rho_{raw}(t)$ is the density matrix constructed from the correlation counts of Eq. (4), together with the effect of the dark counts; and $\mathbb{I}/4$ is the density matrix of an uncorrelated pair of photons, $1/4[|HH\rangle\langle HH| + |HV\rangle\langle HV| + |VH\rangle\langle VH| + |VV\rangle\langle VV|]$.

Figure 7c shows the calculated $HH + VV$ correlations obtained from Eq. (4), as well as $(RL + LR) - (RR + LL)$ correlations indicating the oscillation between $|\Psi\rangle$ and $|\Phi\rangle$, which shows a similar trend to the experimental results as shown in **Figure 7a**. In the next step, we plot the evolution of the calculated concurrence during the exciton lifetime in **Figure 7d**, based on the density matrix constructed from Eq. (5). The measured values are plotted as light green circles, similar to **Figure 7b**, and the results from the simulation are plotted as a solid red line. Surprisingly, it is with great precision that the two data sets agree. The results from the simulation follow the same trend as the measured values, with three regimes:

(I) top, (II) flat, and (III) roll-off. Initially, as the detectors' response function $g(t)$ detects more photons, the plot shows an increase in the concurrence reaching a maximum. However, due to the low timing resolution of the detectors after a period of time, the phase averaging of Eq. (2) becomes significant, and the measured concurrence drops. Once the response function has been fully covered, the phase averaging becomes constant, hence reaching the "flat" part. Finally, as the probability of the exciton emission drops exponentially, the effect of dark counts dominates the actual photon counts, and the tomography system will detect uncorrelated false correlations from the dark counts, the part considered as "roll-off" whereby the concurrence drops further. In addition to this close agreement, the count-weighted average concurrence of the simulation yields $\overline{C}_{\rho_{sim}} = 0.61 \pm 0.01$, which agrees extremely well with the earlier mentioned value from the measurement, $\overline{C}_{\rho} = 0.62 \pm 0.03$, within the error. Thus, the behavior of the two-photon quantum state can be explained by a model which assumes no dephasing, only considering the general features of the source, and the detection systems timing resolution. This indicates that the source at hand is not affected by dephasing during the the exciton decay time once excited at the donor/acceptor level. Hence, this excitation scheme is named "quasi-resonant," as it shows a dephasing-free two-photon quantum state, during the exciton's decay time, without being excited resonantly.

In stark contrast, under non-resonant excitation at the wurtzite InP bandgap, conducting two-photon quantum state tomography reveals the detrimental effect of the surrounding charge noise on the entangled state. By comparing **Figure 8a** and **b**, it becomes clear that shortly after the excitation laser moves to the InP bandgap, the detrimental effects of the excessive charge carriers become evident, ≈ 0.5ns, after the XX emission. Interestingly, these results indicate that during the exciton lifetime, interaction of the exciton state with the charge carriers is the main source of dephasing, not the presence of large nuclear spins, as was generally believed in the community; a finding which is in agreement with a previous work [65], wherein an indium-rich QD was shown not to be affected by the nuclear spins during an exciton lifetime of ≈ 2.5ns.

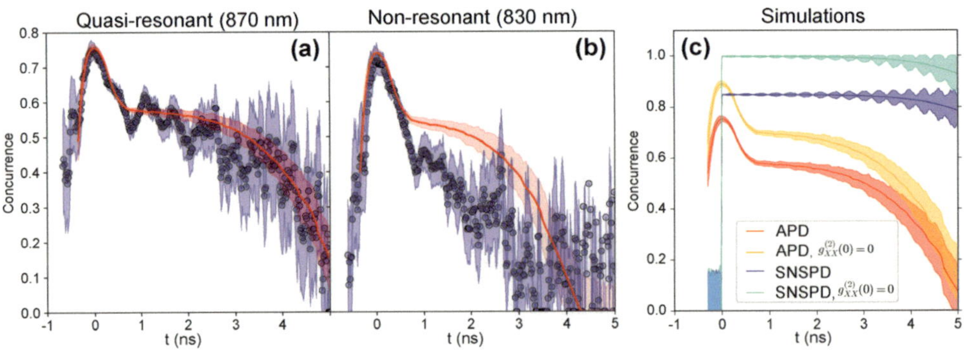

Figure 8.
Effect of the excitation scheme and detection system. (a) Comparison of the theoretical model and results from quasi-resonant excitation indicate suppression of dephasing during the X decay time. (b) Off-resonant excitation at the wurtzite InP bandgap, leads to the mobility of charge carriers and dephasing of the two-photon quantum state shortly after the XX's emission. (c) A combination of two different excitation schemes and detection systems were used to produce the four curves: quasi-resonant excitation and avalanche photodiodes (APDs) (red), resonant TPE and APDs (yellow), quasi-resonant excitation and superconducting nanowire single-photon detectors (SNSPDs) (blue), and resonant TPE and SNSPDs (cyan). Imperfect $g^{(2)}_{XX}(0)$ values in the case of quasi-resonant excitation (red and blue curves), as well as low timing resolution and relatively high noise level of APDs (red and yellow curves), result in the deterioration of the measured concurrence. Impressively, with the application of resonant TPE, and SNSPDs with a timing resolution of $\tau_d \sim 30$ps, and noise level of ~ 1Hz, the detection of perfect entanglement is expected.

As mentioned earlier, the drop observed in the measured concurrence is the result of the low timing resolution of the detectors. Therefore, it is expected that once the detection system is improved, an enhancement in the measured concurrence will be observed. **Figure 8c** shows the result of a simulation when the features of the detection system and/or the excitation scheme have changed. The red curve shows the actual system at hand, quasi-resonant excitation, with $g^{(2)}_{XX}(0) = 0.1$, together with a regular avalanche photodiode (APD) detection system, with a timing resolution of $\tau_d \sim 300$ ps, with τ_d being the FWHM of the response function $g(t)$, and with a dark count rate of $DC \sim 30$ Hz. Upon resonant excitation, the yellow curve, the multiphoton emission of the source is expected to vanish, $g^{(2)}_{XX}(0) = 0.0$, and the uncorrelated photons will not enter the analysis, hence, an expected increase in the concurrence value. However, the general shape of the graph does not change. In the case of conducting the experiment with a fast, low-noise detector, such as with superconducting nanowire single-photon detectors (SNSPDs), with a timing resolution of $\tau_d \sim 30$ ps and a dark count rate of $DC \sim 1$ Hz, the blue curve, not only will the measured value of the concurrence be enhanced, but the overall shape of the graph will change. The drop in concurrence, observed in the case of APDs, will vanish, and the graph will only consist of the "flat" and "roll-off" parts. However the blue curve still suffers from a non-zero multiphoton emission. But, once the source is excited resonantly, and SNSPDs are used, the cyan curve, remarkably, one expects to measure near-unity concurrence. The count-weighted average concurrence in the latter case is $\overline{C} = 0.996 \pm 0.008$.

The way in which the curve of concurrence vs. time is affected by the detectors' response function $g(t)$ can be analyzed in two equivalent ways. In the first approach, a low timing resolution will result in averaging of the relative phase between the two terms in Eq. (2) and turning the pure two-photon quantum state into a mixed state. In this view, at each particular time t, the coincidence counts in the range $[t - \tau_d/2, t + \tau_d/2]$ will be included in the analysis. By dividing this range in shorter time intervals, one can write the resultant measured density matrix, ρ_m, as:

$$\rho_m(t) = \int_{t-\tau_d/2}^{t+\tau_d/2} n(t, \tau_x)\rho(t)dt \tag{6}$$

where $\rho(t) = |\Psi(t)\rangle\langle\Psi(t)|$ and $n(t, \tau_X)$ is the probability of the state being in the state $\rho(t)$, based on an exponential decay. The rate of change in the density matrix is proportional to the *FSS*, since *FSS* results in precession of the state between $|\Psi\rangle$ and $|\Phi\rangle$ according to Eq. (2). Therefore, in the case of a large *FSS*, and a low timing resolution, the measured density matrix will be a result of a mixture of different states in the time interval $[t - \tau_d/2, t + \tau_d/2]$. It was found that the time window for tomography analysis can be chosen to be smaller than τ_d. In the measurements presented here concerning time windows $\Delta t < 100$ps, the concurrence value does not show any change; however, increasing the length of time window above 100 ps reduces the calculated concurrence. This means that the effective time window is slightly less than the FWHM of $g(t)$. It is straightforward to see that as the detectors' timing resolution is enhanced, $\tau_d \rightarrow 0$, the measured density matrix gets closer and closer to the density matrix of a pure state at each instant of time, hence, an increase in the measured concurrence.

In the alternate approach, the uncertainty in timing of the arrival of the photons can be interpreted as an uncertainty in measuring the energy of XX and X, according to $\Delta\tau_d\Delta E = \hbar/2$. In this picture, the timing resolution of the detectors competes with the which-path information introduced by the presence of *FSS* (**Figure 9**). Because of this, a detector with low timing resolution ($\tau_d \gg \hbar/\delta$) will

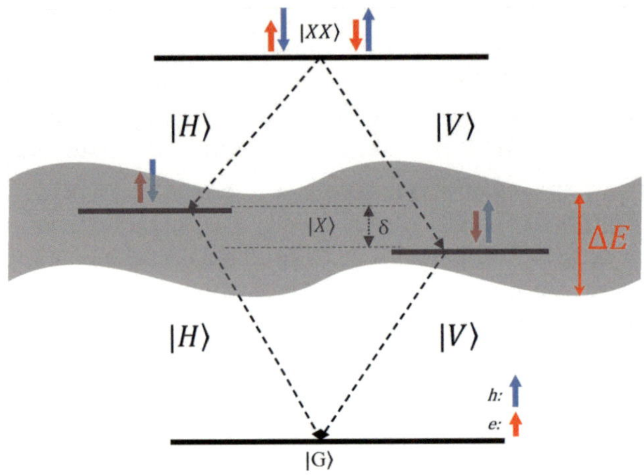

Figure 9.
Detectors' timing resolution and energy uncertainty. The detectors' timing resolution, τ_d, directly leads to an uncertainty in the energy of photons, $\Delta E \tau_d \sim \hbar/2$. For the case of a fast detector, $\tau_d \ll \hbar/\delta$, this uncertainty can smear out the energy difference between the two decay paths and hence retrieve the entanglement, whereas a slow detector, $\tau_d \gg \hbar/\delta$, will push the correlations more toward classical correlations.

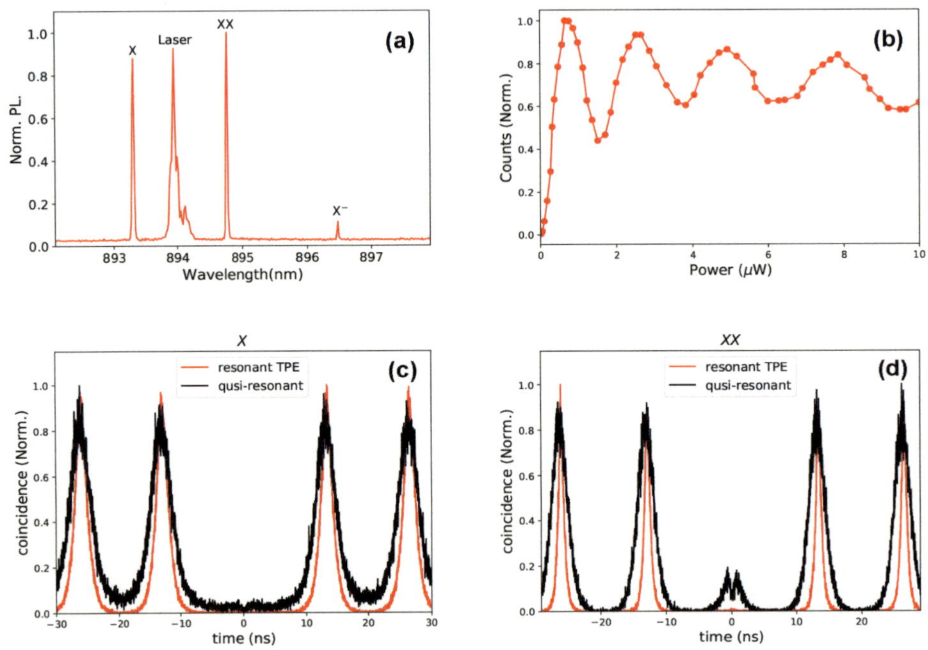

Figure 10.
Resonant two-photon excitation of a nanowire QD. (a) The spectrum of the QD under resonant TPE. The X and XX PL transition rates become more similar as compared to non-resonant excitation, indicating an enhanced pair-production efficiency; and the charged exciton is significantly suppressed, indicating a reduction of excessive charged carriers around the QD. (b) The power-dependent XX count rate exhibits a qualitatively similar Rabi oscillation as the regular direct resonant excitations, qualitatively. (c) and (d) show the comparison between results of $g^{(2)}(0)$ measurements in the case of quasi-resonant and resonant TPE schemes, for X and XX, respectively. Implementation of resonant TPE significantly reduces the emission time jitter of the two states, as well as multiphoton emission of the XX state.

"notice" the energy difference between the two paths, thus degrading the entanglement fidelity. In contrast, a fast detector ($\tau_d \ll \hbar/\delta$) will render the two paths indistinguishable, since the uncertainty in energy will be more than the energy difference of the two paths. In this latter case, the entanglement will be retained and will reach unity when the detector is fast enough to fully erase the which-path information caused by *FSS*.

2.4 Resonant two-photon excitation

In an attempt to realize on-demand entanglement, we have performed performed resonant two-photon excitation on the same sample used by Fognini et al. [45]. The spectrum of the source under resonant TPE is given in **Figure 10a**. As it is evident from comparing this spectrum with the spectra under non-resonant excitation shown in **Figure 6**, the abundance of charge carriers surrounding the QD is significantly suppressed, leading to a lower intensity of the X^- line, as compared to X and XX. Moreover, the PL transition rates of XX and X become closer to each other, a fact that shows an enhancement in pair-production efficiency. By integrating the area under the X and XX PL emission lines and calculating their ratio, we have achieved a pair-production efficiency of $\epsilon_p = 93.6\%$. Proper population of the XX state is affected by the center wavelength of the excitation laser, as well as its bandwidth, the length of which can be controlled via a regular 4f pulse shaper. The population of the XX state in resonant TPE shows a qualitatively similar Rabi oscillation as the regular resonant excitation (**Figure 10b**). The center wavelength and bandwidth of the excitation pulse is chosen so that the π pulse shows the highest possible count rate. Based on taking the setup efficiency and the count rate detected at the π pulse into consideration, the pair-extraction efficiency is reported to be $\epsilon_e = 12.55\%$.

Moreover, under resonant TPE, the multiphoton emission is significantly suppressed. **Figure 10c** and **d** show the results of the second-order correlation function performed on the QD once excited at the donor/acceptor levels and under resonant TPE. For resonant TPE, $g^{(2)}_{XX}(0) = 0.0055 \pm 0.0005$ and $g^{(2)}_{X}(0) = 0.0024 \pm 0.0002$, which demonstrates a two order of magnitude improvement in the case of XX, as compared to the values reported for quasi-resonant excitation.

2.5 State-of-the-art entangled photon sources

The impressive potential for nanowire QDs in detecting entangled photon pairs with near-unity entanglement fidelity is illuminated by the results of the resonant two-photon excitation. Notably, we are now at a point where we can make a comparison between SPDC sources and state-of-the-art QDs in different structures, i.e., self-assembled, micropillar cavities, nanowires, etc. As mentioned earlier, the Poissonian nature of photon-pair emission in SPDC sources limits the performance of such sources to extremely low pair-extraction efficiencies. On the other hand, recent advances in QD growth in various photonic structures have resulted in achieving high entanglement fidelity and high pair-extraction efficiencies, simultaneously. Hüber et al. [67] have reported on measuring an entanglement fidelity of $F = 0.978(5)$, from a self-assembled QD by strain-tuning the *FSS* down to zero. This significant result demonstrates an extensive level of improvement as compared to the results gained from the first generation of self-assembled QDs, where the entanglement fidelity was much lower [38, 68]. The results reported by Fognini et al. [45], in conjunction with the results achieved by resonant TPE, equip us with sufficient information to make such a comparison, the ultimate potential of nanowire QDs in regards to both entanglement fidelity and pair-extraction efficiency, with the values reported for other photonic structures mentioned earlier [26, 32, 33].

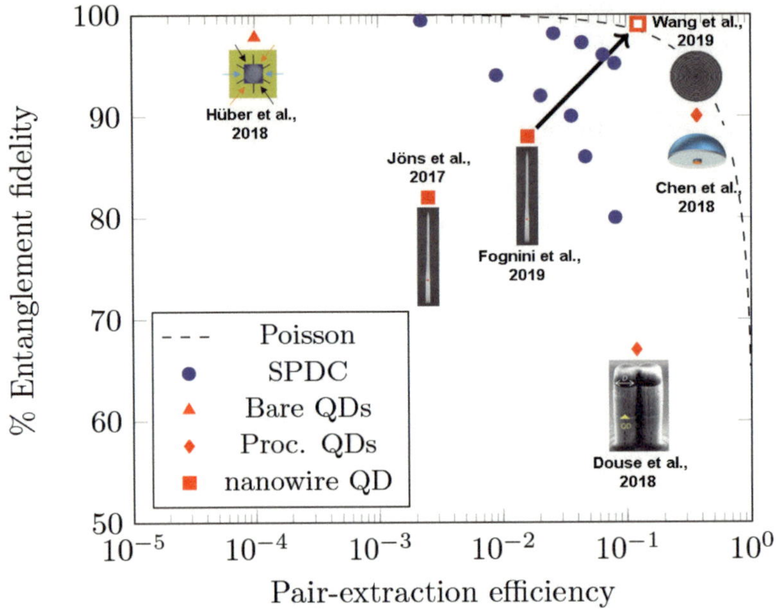

Figure 11.
Performance of state-of-the-art entangled photon sources. Comparison between various quantum light sources in terms of entanglement fidelity and pair-extraction efficiency. Blue circles represent SPDC sources, values taken from [69] and [14]. The dashed line shows the ultimate theoretical limit of such sources, with multiphoton emission probability following a Poisson distribution. The red triangle shows results for a bare self-assembled QD, whereas the red diamonds show the results for QDs in different photonic structures. The red solid squares indicate the values reported for nanowire QDs so far. The analysis performed by Fognini et al. [45] and the results obtained by Ahmadi et al. [66] strongly suggest that the sources used for these two studies have the capacity to surpass the performance of SPDC sources once excited via resonant TPE and measured with a fast, low-noise detector. The graph is adapted and modified from [62].

The result of such a comparison is shown in **Figure 11**. The blue circles show different values reported for entanglement fidelity vs. pair-extraction efficiency for SPDC sources. The values are taken from [69] and [14]. The dashed line shows the theoretical limit of such sources, following a Poisson distribution for the probability of multiphoton emission [70]. The two solid red squares indicate the result of two measurements performed on nanowire QDs by Jöns et al. [62] and Fognini et al. [45]. The latter work shows both an improvement in the measured entanglement fidelity and an improvement in pair-extraction efficiency. Based on the results shown by Fognini et al. [45] and the improvements gained by performing resonant TPE, we can predict measuring near-unity entanglement fidelity once two important modifications are implemented: the resonant TPE scheme is employed, and the detection system is improved to a fast and low-noise one. The final result that we predict by implementing these two changes is shown by the hollow red square. This is an extrapolation of results reported thus far on nanowire QDs based on the enhancement achieved in pair-extraction efficiency and entanglement fidelity, as well as the analysis presented in **Figure 8c**. Therefore, it is confidently predicted that nanowire QDs have the potential to surpass and outperform that of SPDC sources, revealing the significant potential of these sources for quantum communication purposes.

3. Conclusion and discussion

In this chapter, we have given a historical overview of previous methods for attaining pairs of entangled photons from a QD, as well as included the latest research and subsequent recent advances toward enhancement of the performance from such sources. Thus far, several photonic structures have been developed in order to improve the low pair-extraction efficiency of self-assembled QDs, among which bottom-up grown nanowire QDs exhibit considerable promise. Based on the detailed studies of these sources under different excitation schemes along with understanding the effects of detection systems and multiphoton emission on the measured value of entanglement fidelity, we predict nanowire QDs can undoubtedly outperform SPDC sources, once excited via resonant TPE and detected by fast, low-noise detectors.

Admittedly, despite the fact that the results that indicate near-unity fidelity are achievable by nanowire QDs, the finite value of *FSS* will limit the performance of the source once a particular entangled state is required. In this case, a post selection on the collected photons is inevitable, since the two-photon quantum state precesses between the two entangled states Eq. (2). Strain fields [67], electric fields [40], and magnetic fields [68] are the most popular approaches used for addressing this issue; however, two recent proposals seem to be most compatible with a nanowire QD. The first is the method proposed by Fognini et al. [44], which uses a fast-rotating half-wave plate, realized by an electro-optical modulator, in order to change the energy of the photons after they have been emitted and correct for the energy splitting, thus making this approach a universal *FSS* eraser. The second, proposed by Zeeshan et al. [42] on the other hand, corrects the altered symmetry of the exciton's wave function, due to presence of *FSS*, via application of a quadrupole electric field, an approach which requires fabrication of electrical gates in the proximity of the nanowire. In addition to *FSS* tuning, nanowire QDs have been shown to be integrated into designs realizing performance of on-chip optical operations [71]. Such designs enhance light extraction and also allow for developing scalable quantum photonic circuits, paving the way for performing quantum computational processes on a photonic chip [72], using an on-demand entangled photon source.

Excitingly, this research shows that despite the challenges experienced thus far in generating on-demand and optimally entangled photon pairs, the results gained from resonant excitation of a nanowire QD have in fact revealed the enormous potential these sources have to outperform their predecessors. This research and the realization of optimally entangled photon pairs it offers have given quantum foundations, quantum communication, and quantum information a quantum leap forward.

Acknowledgements

The authors gratefully acknowledge the Swiss National Science Foundation, Industry Canada, Natural Sciences and Engineering Research Council of Canada (NSERC) and Transformative Quantum Technologies (TQT), for their funding and support.

Author details

Arash Ahmadi[1]*, Andreas Fognini[2] and Michael E. Reimer[3]

1 Department of Physics and Astronomy, Institute for Quantum Computing, University of Waterloo, Waterloo, Canada

2 Single Quantum, Delft, Netherlands

3 Department of Electrical and Computing Engineering, Institute for Quantum Computing, Waterloo, Canada

*Address all correspondence to: arash.ahmadi@uwaterloo.ca

IntechOpen

© 2020 The Author(s). Licensee IntechOpen. This chapter is distributed under the terms of the Creative Commons Attribution License (http://creativecommons.org/licenses/by/3.0), which permits unrestricted use, distribution, and reproduction in any medium, provided the original work is properly cited.

References

[1] Weston MM, Hall MJW, Palsson MS, Wiseman HM, Pryde GJ. Experimental test of universal complementarity relations. Physical Review Letters. 2013; **110**(22):220402

[2] Shalm LK, Meyer-Scott E, Christensen BG, Bierhorst P, Wayne MA, Stevens MJ, et al. Strong loophole-free test of local realism. Physical Review Letters. 2015;**115**(25):250402

[3] Bouwmeester D, Pan J-W, Mattle K, Eibl M, Weinfurter H, Zeilinger A. Experimental quantum teleportation. Nature. 1997;**390**(6660):575

[4] Pan J-W, Bouwmeester D, Weinfurter H, Zeilinger A. Experimental entanglement swapping: Entangling photons that never interacted. Physical Review Letters. 1998;**80**(18):3891

[5] Ursin R, Tiefenbacher F, Schmitt-Manderbach T, Weier H, Scheidl T, Lindenthal M, et al. Entanglement-based quantum communication over 144 km. Nature Physics. 2007;**3**(7):481

[6] Ekert AK, Rarity JG, Tapster PR, Palma GM. Practical quantum cryptography based on two-photon interferometry. Physical Review Letters. 1992;**69**:1293-1295

[7] Koashi M, Yamamoto T, Imoto N. Probabilistic manipulation of entangled photons. Physical Review A. 2001;**63**: 030301

[8] Qiang X, Zhou X, Wang J, Wilkes CM, Loke T, O'Gara S, et al. Large-scale silicon quantum photonics implementing arbitrary two-qubit processing. Nature Photonics. 2018; **12**(9):534

[9] Burnham DC, Weinberg DL. Observation of simultaneity in parametric production of optical photon pairs. Physical Review Letters. 1970; **25**(2):84

[10] Hong CK, Mandel L. Experimental realization of a localized one-photon state. Physical Review Letters. 1986; **56**(1):58

[11] Shih YH, Sergienko AV, Rubin MH, Kiess TE, Alley CO. Two-photon entanglement in type-II parametric down-conversion. Physical Review A. 1994;**50**(1):23

[12] Verstraete F, Verschelde H. Fidelity of mixed states of two qubits. Physical Review A. 2002;**66**(2):22307

[13] Senellart P, Solomon G, White A. High-performance semiconductor quantum-dot single-photon sources. Nature Nanotechnology. 2017;**12**(11): 1026-1039

[14] Wang X-L, Chen L-K, Wei L, Huang H-L, Liu C, Chao C, et al. Experimental ten-photon entanglement. Physical Review Letters. 2016;**117**(21): 210502

[15] Michler P. Single Quantum Dots: Fundamentals, Applications and New Concepts. Vol. 90. Berlin: Springer Science & Business Media; 2003

[16] Benson O, Santori C, Pelton M, Yamamoto Y. Regulated and entangled photons from a single quantum dot. Physical Review Letters. 2000;**84**(11): 2513-2516

[17] Gywat O, Krenner HJ, Brezovsky J. Spins in Optically Active Quantum Dots. Concepts and Methods. Weinheim: WILEY-VCH, KGaA; 2010

[18] Sipahigil A, Jahnke KD, Rogers LJ, Teraji T, Isoya J, Zibrov AS, et al. Indistinguishable photons from separated silicon-vacancy centers in

diamond. Physical Review Letters. 2014; **113**(11):113602

[19] Sipahigil A, Goldman ML, Togan E, Chu Y, Markham M, Twitchen DJ, et al. Quantum interference of single photons from remote nitrogen-vacancy centers in diamond. Physical Review Letters. 2012;**108**(14):143601

[20] Koperski M, Nogajewski K, Arora A, Cherkez V, Mallet P, Veuillen J-Y, et al. Single photon emitters in exfoliated WSE 2 structures. Nature Nanotechnology. 2015;**10**(6):503

[21] Tran TT, Bray K, Ford MJ, Toth M, Aharonovich I. Quantum emission from hexagonal boron nitride monolayers. Nature Nanotechnology. 2016;**11**(1):37

[22] Aharonovich I, Englund D, Toth M. Solid-state single-photon emitters. Nature Photonics. 2016;**10**(10):631-641

[23] Leonard D, Krishnamurthy M, Reaves CMV, DenBaars SP, Petroff PM. Direct formation of quantum-sized dots from uniform coherent islands of InGaAs on GaAs surfaces. Applied Physics Letters. 1993;**63**(23):3203-3205

[24] Petroff PM, DenBaars SP. MBE and MOCVD growth and properties of self-assembling quantum dot arrays in III-V semiconductor structures. Superlattices and Microstructures. 1994;**15**(1):15

[25] Watanabe K, Koguchi N, Gotoh Y. Fabrication of GaAs quantum dots by modified droplet epitaxy. Japanese Journal of Applied Physics. 2000;**39** (2A):L79

[26] Chen Y, Zopf M, Keil R, Ding F, Schmidt OG. Highly-efficient extraction of entangled photons from quantum dots using a broadband optical antenna. Nature Communications. 2018;**9**(1):1-7

[27] Gérard JM, Sermage B, Gayral B, Legrand B, Costard E, Thierry-Mieg V. Enhanced spontaneous emission by quantum boxes in a monolithic optical microcavity. Physical Review Letters. 1998;**81**(5):1110

[28] Solomon GS, Pelton M, Yamamoto Y. Single-mode spontaneous emission from a single quantum dot in a three-dimensional microcavity. Physical Review Letters. 2001;**86**(17):3903

[29] Gazzano O, De Vasconcellos SM, Arnold C, Nowak A, Galopin E, Sagnes I, et al. Bright solid-state sources of indistinguishable single photons. Nature Communications. 2013;**4**:1425

[30] Somaschi N, Giesz V, De Santis L, Loredo JC, Almeida MP, Hornecker G, et al. Near-optimal single-photon sources in the solid state. Nature Photonics. 2016;**10**(5):340-345

[31] Rodt S, Heitz R, Schliwa A, Sellin RL, Guffarth F, Bimberg D. Repulsive exciton-exciton interaction in quantum dots. Physical Review B. 2003; **68**(3):035331

[32] Dousse A, Suffczyński J, Beveratos A, Krebs O, Lemaître A, Sagnes I, et al. Ultrabright source of entangled photon pairs. Nature. 2010; **466**(7303):217-220

[33] Wang H, Hu H, Chung T-H, Qin J, Yang X, Li J-P, et al. On-demand semiconductor source of entangled photons which simultaneously has high fidelity, efficiency, and indistinguishability. Physical Review Letters. 2019; **122**:113602

[34] Bayer M, Ortner G, Stern O, Kuther A, Gorbunov AA, Forchel A, et al. Fine structure of neutral and charged excitons in self-assembled In(Ga)As/(Al)GaAs quantum dots. Physical Review B. 2002;**65**(19):195315

[35] Singh R. Tuning fine structure splitting and exciton emission energy in semiconductor quantum dots. Journal of

Luminescence. February 2018;**202**: 118-131

[36] Stevenson RM, Thompson RM, Shields AJ, Farrer I, Kardynal BE, Ritchie DA, et al. Quantum dots as a photon source for passive quantum key encoding. Physical Review B. 2002; **66**(8):081302

[37] Santori C, Fattal D, Pelton M, Solomon GS, Yamamoto Y. Polarization-correlated photon pairs from a single quantum dot. Physical Review B. 2002;**66**(4):045308

[38] Young RJ, Stevenson RM, Atkinson P, Cooper K, Ritchie DA, Shields AJ. Improved fidelity of triggered entangled photons from single quantum dots. New Journal of Physics. 2006;**8**(2):29

[39] Akopian N, Lindner NH, Poem E, Berlatzky Y, Avron J, Gershoni D, et al. Entangled photon pairs from semiconductor quantum dots. Physical Review Letters. 2006;**96**(13):130501

[40] Kowalik K, Krebs O, Lemaître A, Laurent S, Senellart P, Voisin P, et al. Influence of an in-plane electric field on exciton fine structure in InAs-GaAs self-assembled quantum dots. Applied Physics Letters. 2005;**86**(4):041907

[41] Seidl S, Kroner M, Högele A, Karrai K, Warburton RJ, Badolato A, et al. Effect of uniaxial stress on excitons in a self-assembled quantum dot. Applied Physics Letters. 2006;**88**(20): 203113

[42] Zeeshan M, Sherlekar N, Ahmadi A, Williams RL, Reimer ME. Proposed scheme to generate bright entangled photon pairs by application of a quadrupole field to a single quantum dot. Physical Review Letters. 2019; **122**(22):227401

[43] Trotta R, Zallo E, Ortix C, Atkinson P, Plumhof JD, Van den Brink J, et al. Universal recovery of the energy-level degeneracy of bright excitons in InGaAs quantum dots without a structure symmetry. Physical Review Letters. 2012;**109**(14):147401

[44] Fognini A, Ahmadi A, Daley SJ, Reimer ME, Zwiller V. Universal fine structure eraser for quantum dots. Optics Express. 2017;**26**(19):25-28

[45] Fognini A, Ahmadi A, Zeeshan M, Fokkens JT, Gibson SJ, Sherlekar N, et al. Dephasing free photon entanglement with a quantum dot. ACS Photonics. 2019;**6**(7):1656-1663

[46] Reimer ME, Bulgarini G, Fognini A, Heeres RW, Witek BJ, Versteegh MAM, et al. Overcoming power broadening of the quantum dot emission in a pure wurtzite nanowire. Physical Review B. 2016;**93**(19):1-9

[47] Flagg EB, Polyakov SV, Thomay T, Solomon GS. Dynamics of nonclassical light from a single solid-state quantum emitter. Physical Review Letters. 2012; **109**(16):1-5

[48] Michler P. Quantum Dots for Quantum Information Technologies. Vol. 237. Berlin: Springer; 2017

[49] Brunner K, Abstreiter G, Böhm G, Tränkle G, Weimann G. Sharp-line photoluminescence and two-photon absorption of zero-dimensional biexcitons in a GaAs/AlGaAs structure. Physical Review Letters. 1994;**73**: 1138-1141

[50] Flissikowski T, Betke A, Akimov IA, Henneberger F. Two-photon coherent control of a single quantum dot. Physical Review Letters. 2004;**92**: 227401

[51] Stufler S, Machnikowski P, Ester P, Bichler M, Axt VM, Kuhn T, et al. Two-photon Rabi oscillations in a single Inx Ga1-x As GaAs quantum dot. Physical

Review B: Condensed Matter and Materials Physics. 2006;**73**(12):1-7

[52] Jayakumar H, Predojević A, Huber T, Kauten T, Solomon GS, Weihs G. Deterministic photon pairs and coherent optical control of a single quantum dot. Physical Review Letters. 2013;**110**:135505

[53] Müller M, Bounouar S, Jöns KD, Glässl M, Michler P. On-demand generation of indistinguishable polarization-entangled photon pairs. Nature Photonics. 2014;**8**(3):224-228

[54] Schweickert L, Jöns KD, Zeuner KD, Da Silva SFC, Huang H, Lettner T, et al. On-demand generation of background-free single photons from a solid-state source. Applied Physics Letters. 2018; **112**(9):1-4

[55] Claudon J, Bleuse J, Malik NS, Bazin M, Jaffrennou P, Gregersen N, et al. A highly efficient single-photon source based on a quantum dot in a photonic nanowire. Nature Photonics. 2010;**4**(3):174

[56] Gregersen N, Nielsen TR, Claudon J, Gérard J-M, Mørk J. Controlling the emission profile of a nanowire with a conical taper. Optics Letters. 2008; **33**(15):1693-1695

[57] Reimer ME, Bulgarini G, Akopian N, Hocevar M, Bavinck MB, Verheijen MA, et al. Bright single-photon sources in bottom-up tailored nanowires. Nature Communications. 2012;**3**:737

[58] Bulgarini G, Reimer ME, Bavinck MB, Jons KD, Dalacu D, Poole PJ, et al. Nanowire waveguides launching single photons in a gaussian mode for ideal fiber coupling. Nano Letters. 2014;**14**(7):4102-4106

[59] Brown RH, Twiss RQ. A test of a new type of stellar interferometer on sirius. Nature. 1956;**178**(4541): 1046-1048

[60] James DFV, Kwiat PG, Munro WJ, White AG. Measurement of qubits. Physical Review A. 2001;**64**:052312

[61] Versteegh MAM, Reimer ME, Jöns KD, Dalacu D, Poole PJ, Gulinatti A, et al. Observation of strongly entangled photon pairs from a nanowire quantum dot. Nature Communications. 2014;**5**:5298

[62] Jöns KD, Schweickert L, Versteegh MAM, Dalacu D, Poole PJ, Gulinatti A, et al. Bright nanoscale source of deterministic entangled photon pairs violating Bell's inequality. Scientific Reports. 2017;**7**(1):1-11

[63] Clauser JF, Horne MA, Shimony A, Holt RA. Proposed experiment to test local hidden-variable theories. Physical Review Letters. 1969;**23**(15):880

[64] Bennett CH, DiVincenzo DP, Smolin JA, Wootters WK. Mixed-state entanglement and quantum error correction. Physical Review A: Atomic, Molecular, and Optical Physics. 1996; **54**(5):3824

[65] Sénés M, Liu BL, Marie X, Amand T, Gérard JM. Spin Dynamics of Neutral and Charged Excitons in InAs/GaAs Quantum Dots. Dordrecht: Springer Netherlands; 2003. pp. 79-88

[66] Ahmadi A, Mastrovich M, Hosseini S, Jöns KD, Reimer ME. Resonant excitation of a quantum dot in a photonic nanowire. 2020;**5**:3824. Submission in process

[67] Huber D, Reindl M, Filipe S, Schimpf C, Martín-sánchez J, Huang H, et al. Strain-tunable GaAs quantum dot: A nearly dephasing-free source of entangled photon pairs on demand. Physical Review Letters. 2018;**121**(3): 33902

[68] Stevenson RM, Young RJ, Atkinson P, Cooper K, Ritchie DA, Shields AJ. A semiconductor source of

triggered entangled photon pairs. Nature. 2006;**439**(7073):179-182

[69] Scarani V, de Riedmatten H, Marcikic I, Zbinden H, Gisin N. Four-photon correction in two-photon bell experiments. European Physical Journal D: Atomic, Molecular, Optical and Plasma Physics. 2005;**32**(1):129-138

[70] Daley S. Electro-Optic Rotating Half-Waveplate for a Quantum Dot Fine-Structure Eraser. Ontario: UWSpace, University of Waterloo; 2019

[71] Elshaari AW, Zadeh IE, Fognini A, Reimer ME, Dalacu D, Poole PJ, et al. On-chip single photon filtering and multiplexing in hybrid quantum photonic circuits. Nature Communications. 2017;**8**(1):379

[72] Ma X, Fung C-HF, Lo H-K. Quantum key distribution with entangled photon sources. Physical Review A. 2007;**76**:012307

Chapter 4

Interactions of Positrons and Electrons with Hydrogenic Systems, Excitation, Resonances, and Photoabsorption in Two-Electron Systems

Anand K. Bhatia

Abstract

There are a number of approaches to study interactions of positrons and electrons with hydrogenic targets. Among the most commonly used are the method of polarized orbital, the close-coupling approximation, and the *R*-matrix formulation. The last two approaches take into account the short-range and long-range correlations. The method of polarized orbital takes into account only long-range correlations but is not variationally correct. This method has recently been modified to take into account both types of correlations and is variationally correct. It has been applied to calculate phase shifts of scattering from hydrogenic systems like H, He^{+}, and Li^{2+}. The phase shifts obtained using this method have lower bounds to the exact phase shifts and agree with those obtained using other approaches. This approach has also been applied to calculate resonance parameters in two-electron systems obtaining results which agree with those obtained using the Feshbach projection-operator formalism. Furthermore this method has been employed to calculate photodetachment and photoionization of two electron systems, obtaining very accurate cross sections which agree with the experimental results. Photodetachment cross sections are particularly useful in the study of the opacity of the sun. Recently, excitation of the atomic hydrogen by electron impact and also by positron impact has been studied by this method.

Keywords: scattering, resonances, photoabsorption, excitation

1. Introduction

The discovery of an electron by J.J. Thomson in 1897 led to the development of physics beyond the classical physics. Proton was discovered by Rutherford in 1909. Niels Bohr proposed a model of the structure of hydrogen atoms in 1913. Neutron was discovered by Chadwick in 1932. Other important discoveries were of X-rays and radioactivity in 1896. In 1926, Erwin Schrödinger formulated an equation to determine the wave function of quantum mechanical system. According to Max Born, the wave function can be interpreted as a probability of finding a particle at a specific point in space and time. Many processes could be studied due to such developments in physics. For example, an incoming wave behaves like a particle in processes like Compton scattering and photoabsorption. Particularly, Geiger and Bothe, using coincident counters, showed that the time between the arrival of the incident wave and the motion of the electron is of the order of 10^{-7} second. If the incident wave acted as a wave, the time would have been much longer. Also the experiment of Compton and Simon showed that energy is conserved at every point of the scattering process.

We discuss here scattering of electrons by hydrogenic systems since the wave function of the target is known exactly, and therefore we can test various theories or approximations. When the target consists of more than one electron, a reasonably accurate wave function can only be written using various configurations of the target (called configuration interaction approximation). Among the various approximations for scattering are the exchange approximation [1], the Kohn variational method [2], and the method of polarized orbitals [3] which takes into account the polarization of the target due to the incident electron. The incident electron creates an electric field which results in a change of energy of the target given by $\Delta E = -\frac{1}{2}\alpha E^2$, where α is the polarizabilty of the target and is equal to 4.5 a_0^3 in the case of a hydrogen target. The polarization is possible only when the incident electron is outside the target, according to the method of polarized orbitals [3]. However, this method includes only the long-range $-1/r^4$ potential and not the short-range correlations, and the method is not variationally correct. It is possible to use the Feshbach formalism [4] to take into account the short-range correlations via an optical potential. This method gives rigorous lower bonds on the phase shifts, i.e., they are lower than the exact phase shifts. However, it is difficult to include the long-range correlations at the same time. The close-coupling approach takes into account both the long-range and short-range correlations [5] and is variationally correct. However, a large number of target states must be included to obtain converged results and the correct polarizability. A Feshbach resonance below the $n = 2$ threshold of hydrogen atom was first discovered in a close-coupling calculation. A method which has been applied extensively to atomic, molecular, and nuclear physics is the R-matrix method introduced by Wigner and Eisenbud in 1947. In this method, configuration space is divided in such a way that all correlations are described within a radius $r = a$, and outside this radius simple continuum functions can be used while matching the inside and outside functions at the boundary $r = a$. This method was introduced in atomic physics by Burke. A method called the hybrid theory [6] has been introduced in which the short-range and long-range correlations are taken into account at the same time and the polarization takes place whether the incident electron is outside or inside the target. This method is variationally correct. The equations for the scattering function are very detailed and are given in [6].

2. Scattering function calculations

In the exchange approximation [1], we write the wave function of incident electron and the target as

$$\Psi\left(\vec{r}_1, \vec{r}_2\right) = u\left(\vec{r}_1\right)\phi_0\left(\vec{r}_2\right) \pm (1 \leftrightarrow 2). \tag{1}$$

In the above equation, the plus sign refers to the singlet state, and the minus sign refers to the triplet state, $u\left(\vec{r}_1\right)$ is the scattering function, and

$$\phi_0\left(\vec{r}_2\right) = 2\sqrt{Z^3}e^{-Zr_2}Y_{00}(\hat{r}_2) \tag{2}$$

is the target wave function. The scattering wave function of the incident electron is obtained from

$$\int \left[\phi_0\left(\vec{r}_2\right)|H - E|\Psi\right] d\vec{r}_2 = 0 \tag{3}$$

where H is the Hamiltonian and E is the energy of e-target system. In Rydberg units,

$$H = -\nabla_1^2 - \nabla_2^2 - \frac{2Z}{r} - \frac{2Z}{r_2} - \frac{2}{r_{12}}, \tag{4}$$

$$E = k^2 - Z^2, \tag{5}$$

where k^2 is the kinetic energy of the incident electron and Z is the nuclear charge of the target, assumed fixed. The scattering equation is

$$\begin{aligned}&\left[\frac{d^2}{dr_1^2} - \frac{l(l+1)}{r_1^2} + 2e^{-2Zr_1}\left(1 + \frac{1}{r_1}\right) + k^2\right] u_l(r_1)\\ &\pm 4Z^3 e^{-Zr_1}\left[(Z^2 + k^2)\delta_{l0} r_1 \int_0^\infty e^{-Zr_2} r_2 u_l(r_2) dr_2 - \frac{2}{(2l+1)} y_l(u_l, \phi_0)\right]\\ &= 0\end{aligned} \tag{6}$$

where

$$y_l(\phi_0, u_l) = \frac{1}{r^{l+1}} \int_0^r r_2^l \phi_0(r_2) u_l(r_2) dr_2 + r^l \int_r^\infty \frac{\phi_0(r_2) u_l(r_2)}{r^{l+1}} dr_2 \tag{7}$$

Phase shift η (radians) of a partial wave l is calculated from the asymptotic value of the scattering function.

$$\lim_{r\to\infty} u(r) = \sin\left(kr - l\frac{\pi}{2} + \eta\right). \tag{8}$$

In hybrid theory [6], we replace Eq. (1) by

$$\Psi_L\left(\vec{r}_1, \vec{r}_2\right) = u_L\left(\vec{r}_1\right)\Phi^{pol}\left(\vec{r}_1, \vec{r}_2\right) \pm (1 \leftrightarrow 2) + \Phi_L\left(\vec{r}_1, \vec{r}_2\right), \tag{9}$$

where the polarized target function is given in [7] and is defined as

$$\Phi^{pol}\left(\vec{r}_1, \vec{r}_2\right) = \phi_0\left(\vec{r}_2\right) - \frac{\chi(r_1)}{r_1^2}\frac{u_{1s\to p}(r_2)}{r_2}\frac{\cos(\theta_{12})}{\sqrt{\pi Z}}. \tag{10}$$

The cutoff function, instead of $\varepsilon(r_1, r_2) = 1$ for $r_1 > r_2$, is used in the electron-hydrogen scattering calculation [7], and a cutoff function $\chi(r_1)$ given by Shertzer and Temkin [8] of the form

$$\chi_{ST}(r_1) = 1 - e^{-2Zr_1}\left[\frac{(Zr_1)^4}{3} + \frac{4(Zr_1)^3}{3} + 2(Zr_1)^2 + 2Zr_1 + 1\right] \tag{11}$$

is used in this calculation. It can also be of the form

$$\chi(r_1) = \left(1 - e^{-\beta r_1}\right)^n, \tag{12}$$

where the exponent $n \geq 3$ and β, another parameter which is a function of k, can be used to optimize the phase shifts. Both cutoffs (11) and (12) are used in calculations on scattering. $\Phi_L(r_1, r_2)$ is a short-range correlation function [6] which can be written using the Euler angle decomposition [9] for all partial waves. This formulism allows the separation of the angular parts and the radial parts consisting of radial coordinates r_1, r_2, and r_{12}. This facilitates writing the equation for radial functions [9]. The function $u_{1s\to p}(r_1)$ in Eq. (10) is given by

$$u_{1s\to p}(r_2) = e^{-Zr_2}\left(\frac{Z}{2}r_2^3 + r_2^2\right) \tag{13}$$

A comparison of results for singlet and triplet phase shifts obtained using different methods is given in **Tables 1** and **2**. Results from most methods agree. A comparison of the singlet and triplet phase shifts obtained by the R-matrix and the hybrid theory is given in **Figure 1**. The curves for the two methods cannot be distinguished, showing that accurate result can be obtained using the hybrid theory [6].

The scattering length a is defined by.

$$\lim_{k\to 0} k \cot \eta = -1/a. \tag{14}$$

The scattering length is calculated at a distance R, and there is a correction for the long-range polarization [12].

$$a = a(R) - \alpha\left(\frac{1}{R} - \frac{\alpha}{R^2} + \frac{\alpha^2}{3R^3}\right), \tag{15}$$

where a is the true scattering length and α is the polarizability of the hydrogen atom. At R = 117.3088, ^1S scattering length $a(R)$ is 5.96554, obtained using χ_β. The corrected ^1S scattering length is 5.96611. At R = 349.0831, the ^3S scattering length is 1.781542, obtained using χ_β, and it is 1.76815, after correction for the long-range polarization given by [12]. The ^1S and ^3S scattering lengths are therefore 5.96611 and

k	EA[a]	PO[b]	Kohn[c]	Close-coupling[d]	R-matrix[e]	Feshbach method[f]	Hybrid theory[g]
0.1	2.396	2.583	2.553	2.491	2.550	2.55358	2.55372
0.2	1.870	2.144	2.673	1.9742	2.062	2.06678	2.06699
0.3	1.508	1.750	1.6964	1.519	1.691	1.09816	1.69853
0.4	1.239	1.469	1.4146	1.257	1.410	1.41540	1.41561
0.5	1.031	1.251	1.202	1.082	1.196	1.20094	1.20112
0.6	0.869		1.041		1.035	1.04083	1.04110
0.7	0.744	0.947	0.930		0.925	0.93111	0.93094
0.8	0.651	0.854	0.886	0.608		0.88718	0.88768

[a]*Ref. [1].*
[b]*Ref. [7].*
[c]*Ref. [2].*
[d]*Ref. [5].*
[e]*Ref. [10].*
[f]*Ref. [11].*
[g]*Ref. [6].*

Table 1.
Comparison of ^{1}S phase shifts obtained in different methods.

k	EA[a]	PO[b]	Kohn[c]	Close-coupling[d]	R-matrix[e]	Feshbach method[f]	Hybrid theory[g]
0.1	2.908	2.949	2.9388	2.9355	2.939	2.93853	2.93856
0.2	2.679	2.732	2.7171	2.715	2.717	2.71741	2.71751
0.3	2.461	2.519	2.4996	2.461	2.500	2.49975	2.49987
0.4	2.257	2.320	2.2938	2.2575	2.294	2.29408	2.29465
0.5	2.070	2.133	2.1046	2.0956	2.105	2.10454	2.10544
0.6	1.901		1.9329		1.933	1.93272	1.93322
0.7	1.749	1.815	1.7797		1.780	1.77950	1.77998
0.8	1.614	1.682	1.643	1.616		1.64379	1.64425

[a]*Ref. [1].*
[b]*Ref. [7].*
[c]*Ref. [2].*
[d]*Ref. [5].*
[e]*Ref. [10].*
[f]*Ref. [11].*
[g]*Ref. [6].*

Table 2.
Comparison of ^{3}S *phase shifts obtained in different methods.*

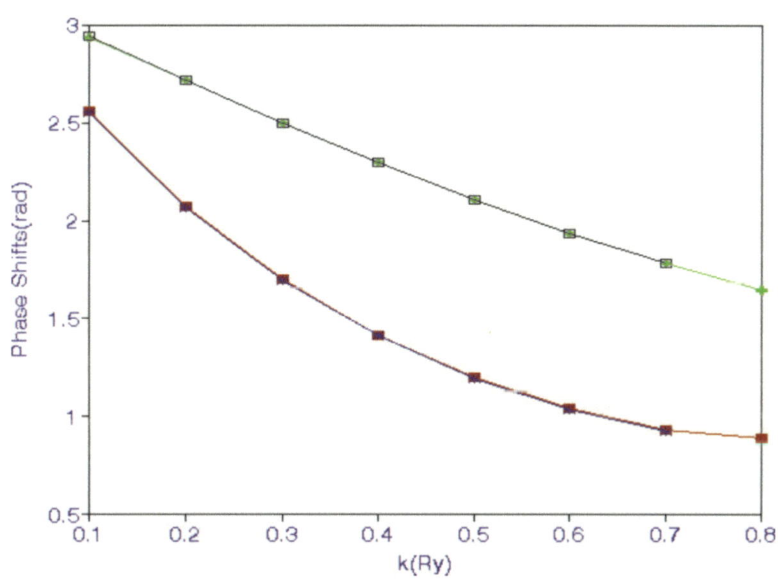

Figure 1.
The upper represents singlet phase shifts obtained in the hybrid and R-*matrix theories. The two curves cannot be distinguished. Similarly, the lower curves represent triplet phase shifts obtained in the two theories.*

1.76815, which agree with those calculated by Schwartz [2], using the Kohn variational method. In this method, scattering lengths have upper bounds, while phase shifts do not have any bounds in this method (Kohn variational method). Similar calculations for the phase shifts for e-He^+ and e-Li^{2+} have been carried out in Ref. [13].

Resonance parameters in two-electron systems have been calculated using various approaches. Among them are the stabilization method, the complex-rotation method, the close-coupling method, and the Feshbach projection-operator formalism. In the hybrid theory, phase shifts have been calculated in the resonance region [13] and are fitted to the Breit-Wigner form

$$\eta_{calc.}(E) = \eta_0 + AE + \tan^{-1}\frac{0.5\Gamma}{(E_R - E)}. \tag{16}$$

In the above equation, $E = k^2$ is the incident energy, $\eta_{calc.}$ are the calculated phase shifts, and η_0, A, Γ, and E_R are the fitting parameters. E_R and Γ represent the resonance position and resonance width, respectively. We find that in the hybrid theory, the He singlet resonance is at E_R = 57.8481 eV with respect to the ground state of He and Γ = 0.1233 eV. They agree well with E_R = 57.8435 eV and Γ = 0.125 eV obtained using the Feshbach projection-operator formalism [14]. A similar calculation [13] has been carried out for Li^{2+} resonance parameters. The resonance parameters agree with those obtained in [14].

P-wave phase shifts have been calculated for scattering of electrons from He^+ and Li^{2+} in Ref. [15] and in Ref. [16], respectively. Singlet P and triplet P phase shifts are shown in **Table 3** for e + He^+ and for e + Li^{2+} scattering. Phase shifts for e + He^+ agree well with those obtained by Oza [17] using the close-coupling approximation. Phase shifts for e + Li^{2+} agree with those obtained by Gien [18] using the Harris-Nesbet method.

k	^{1}P	^{3}P	^{1}P	^{3}P
	e + He^+ [15]		e + Li^{2+} [16]	
0.1	−0.038308	0.21516	−0.049083	0.16323
0.2	−0.038956	0.21683	−0.048990	0.16334
0.3	−0.039873	0.21945	−0.048934	0.16341
0.4	−0.040902	0.22283	−0.48823	0.16351
0.5	−0.041469	0.22662	−0.048565	0.16360
0.6	−0.041641	0.23088	−0.048306	0.16379
0.7	−0.041438	0.23417	−0.047972	0.16382
0.8	−0.039927	0.23753	−0.047547	0.16374
1.0	−0.037132	0.24205	−0.045966	0.16409
1.1	−0.035430	0.24323	−0.045029	0.16399
1.3	−0.026419	0.24370	−0.042670	0.16345
1.4	−0.020773		−0.041251	0.16299
1.6			−0.037973	0.16158

Table 3.
Phase shifts (radians) for e + He^+ and e + Li^{2+} scattering.

3. Photoabsorption

Photodetachment and photoionization are required to calculate radiative-attachment cross sections. The recombination rates are required to calculate the ionization balance in astrophysical plasmas. Cross sections for bound-free transitions of H^- are required to account for the absorption in the solar atmosphere [19]. The opacity in the sun is due to photodetachment and free-free absorption of the radiation:

$$h\nu + H^- \rightarrow H + e, \tag{17}$$

$$h\nu + e + H \rightarrow e + H, \tag{18}$$

In the first process, after absorption of the radiation by the bound electron, it becomes a free electron in the final state, while in the free-free transition, the electron is in the continuum state in the initial state as well as in the final state. It is possible to have the following reactions which help molecular formation:

$$\begin{aligned} e + H &\rightarrow H^- + h\nu \\ H^- + H &\rightarrow H_2 + e. \end{aligned} \tag{19}$$

The photoabsorption cross section in length form and in units of a_o^2 for the transition from the initial states I to the final state f is given by

$$\sigma\left(a_0^2\right) = 4\pi\alpha\omega| < \Psi_f|z_1 + z_2|\Phi_i > |^2. \tag{20}$$

In the above equation, α is the fine-structure constant, k is the momentum of the outgoing electron, $\omega = I + k^2$ is the energy of the incident photon, and I is the ionization potential of the system absorbing the photon. Photoabsorption cross sections for H^-, He, and Li^+ are given in [16] and in **Table 4**. The correlated wave functions for H^-, He, and Li^+ are of Hylleraas form having terms 364, 220, and 165, respectively. A Hyllerraas wave function is given by

$$\Psi(r_1, r_2, r_{12}) = \sum C_{lmn}\left[e^{-\gamma r_1 - \delta r_2} r_1^l r_2^m r_{12} \pm (1 \leftrightarrow 2)\right]. \tag{21}$$

In Eq. (21), C's are the linear parameters, while γ and δ are nonlinear parameters, and $r_{12} = |\vec{r}_1 - \vec{r}_2|$. Ohmura and Ohmura [20] have calculated the photodetachment cross section of H^-, using the effective range theory and the loosely bound structure of hydrogen ion. These cross sections are shown in **Figure 2**. It should be noticed that with the outgoing energy of the electron going to zero, the cross section goes to zero when the remaining system is neutral and is finite when the remaining system is an ion [16].

A comparison of the cross sections of the ground state of He with those obtained in the R-matrix [21] is shown in **Figure 3**. The agreement is very good except at k = 0.6 where the R-matrix result is slightly lower. It seems that the cross section has not been calculated accurately at this k. The precision measurements of the cross sections of photoionization of He by Samson et al. [22] are also shown in **Figure 3**,

k	H^-	He	Li^+
0.1	15.3024	7.3300	
0.2	38.5443	7.1544	2.5677
0.3	35.2318	6.8716	2.5231
0.4	24.4774	6.4951	2.4373
0.5	16.0858	6.0461	2.3870
0.6	10.7410	5.5925	2.2988
0.7	7.4862	5.0120	2.0005
0.8	5.6512	4.4740	2.0925
0.9		3.9296	1.9792

Table 4.
Photoabsorption cross (Mb) for the ground state of H^-, He, and Li^+.

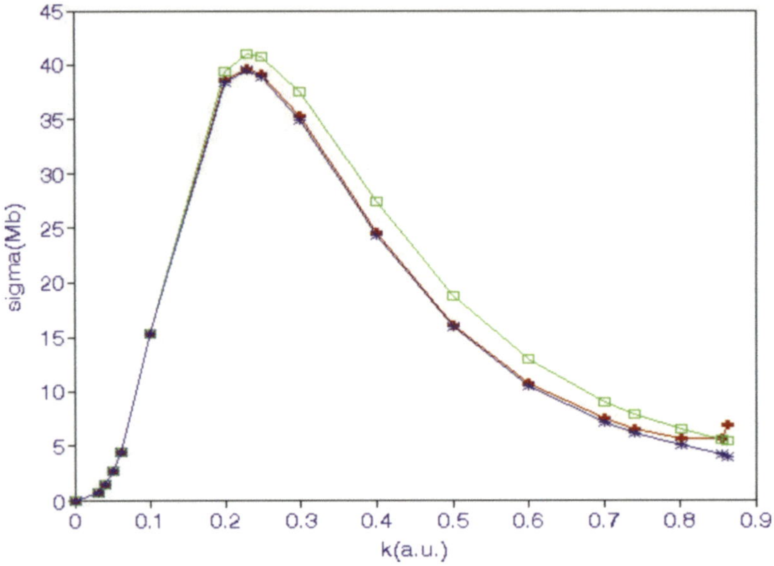

Figure 2.
Photodetachment of a hydrogen ion. The lowest curve is obtained when only the long-range correlations are included; the middle curve is obtained when the short-range and long-range correlations are also included. The top curve is obtained using Ohmura and Ohmura formulation.

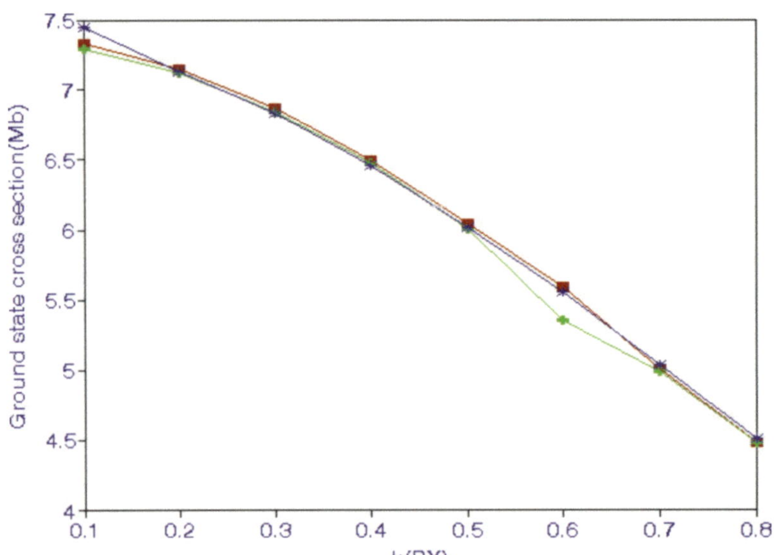

Figure 3.
The upper curve represents photoionization cross sections (Mb) for the ground state of He in the hybrid theory, while the lower represents cross sections obtained in R-matrix calculations. The curve starting at the top left represents experimental results of Samson et al. [22].

showing that the agreement with the cross sections obtained using the hybrid theory is very good.

Similarly, cross sections have been calculated in [16] for the (1s2s) ^1S and (1s2s) ^3S metastable states of He and Li^+. The number of Hylleraas terms used are 455 and 364 for He singlet and triplet states, respectively. For Li^+, 120 and 220 terms are used for the singlet and triplet states, respectively. These cross sections are comparable to those obtained for the ground state. A comparison of photoionization cross

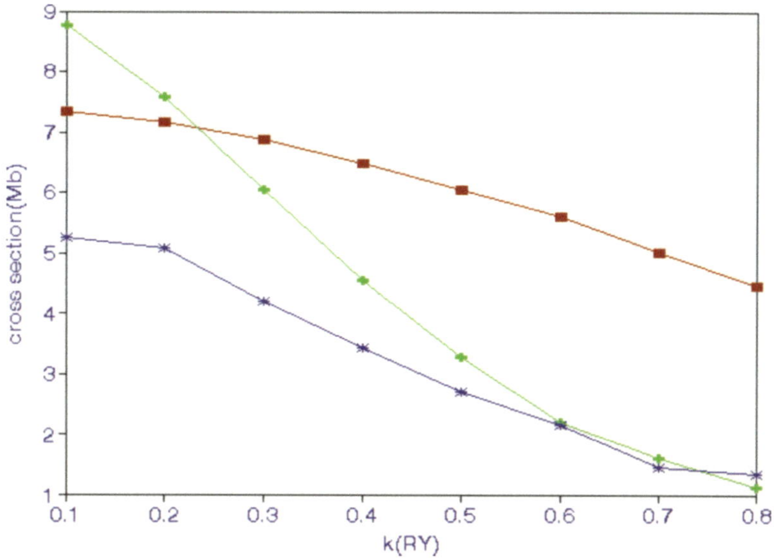

Figure 4.
Photoionization cross section for the ground state (middle curve from the right), for the (1s2s) singlet S state (top curve), and for the (1s2s) triplet S state (lower curve) of the He atom.

sections of the ground state, metastable states singlet S, and triplet S is shown in **Figure 4**. The cross section for the singlet (1s2s) state is highest at $k = 0.1$, and then the cross sections decrease rapidly.

4. Excitation

Excitation of the $1S$ state of atomic hydrogen to the $2S$ state has been calculated using the hybrid theory in the distorted wave approximation [23]. The total excitation cross section is written in the form

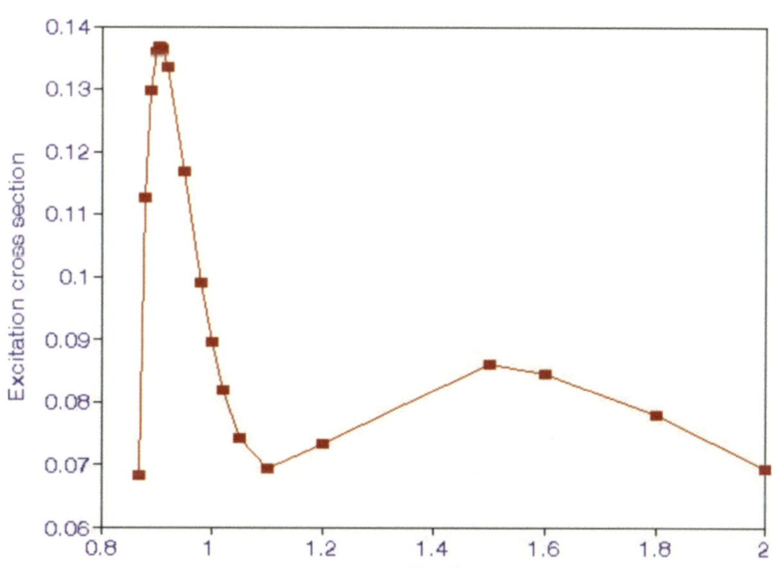

Figure 5.
Cross section (Mb) for exciting the 1S state to 2S state of atomic hydrogen by electron impact.

k	He		Li^+	
	$(1s2s)^1S$	$(1s2s)^3S$	$(1s2s)^1S$	$(1s2s)^3S$
0.1	8.7724	5.2629	2.4335	2.9889
0.2	7.5894	5.0795	2.3742	2.8570
0.3	6.0523	4.2004	2.2287	2.6434
0.4	4.5403	3.4403		2.3733
0.5	3.2766	2.7189		2.0865
0.6	2.2123	2.1531		1.7962
0.7	1.6047	1.4564		1.5182
0.8	1.1230	1.3539		1.2627

Table 5.
Photoionization cross sections (Mb) for the metastable states of He and Li^+ with the outgoing electron with momentum k.

$$\sigma = \frac{k_f}{k_i} \int |T_{fi}|^2 d\Omega. \tag{22}$$

In the above equation, k_i and k_f are the incident and final momenta, and T_{fi} is the matrix element for the excitation of the initial state i to the final state f and is given by

$$T_{fi} = -(1/4\pi) < \Psi_f |V| \Psi_i > . \tag{23}$$

$$V = -\frac{2Z}{r_1} + \frac{2}{r_{12}}. \tag{24}$$

Z is the nuclear charge, and r_1 and r_2 indicate the position of the incident and target electron, respectively. The excitation cross sections at $k(\text{Ry}) = 0.8–2$ are shown in **Figure** 5. There is a maximum at $k = 0.907$ and another at $k = 1.50$ (**Table** 5).

5. Recombination

Recombination rate coefficients for a process like that indicated in Eq. (17) have been calculated in Ref. [16] for the ground states as well as for the metastable states. The attachment cross section σ_a is given by

$$\sigma_a = \left(\frac{h\nu}{cp_e}\right)^2 \frac{g(f)}{g(i)} \sigma. \tag{25}$$

The above relation between the photodetachment and photoionization follows from the principle of detailed balance, where $g(i)$ and $g(f)$ are the weight factors for the initial and final states, and $p_e = k$ is the electron momentum. The attachment cross sections are in general smaller than the photoabsorption cross sections. Recombination rate coefficients, averaged over the Maxwellian velocity distribution, are given in **Table 6** for H^-, He, and Li^+ for a temperature range from 1000 to 40,000 K.

Elastic P-wave scattering from e- Be^{3+}, C^{5+}, and O^{7+} and photoionization in two-electron systems have been carried out. Phase shifts and photoionization cross sections are given in [24] for the ground as well as metastable states. In order to

T	$\alpha_R(T) \times 10^{15}$, H$^-$	$\alpha_{(R)}(T) \times 10^{13}$, He	$\alpha_R(T) \times 10^{13}$, Li$^+$
1000	0.99	2.50	0.12
2000	1.28	2.39	1.04
5000	2.40	1.87	2.62
7000	2.82	1.66	2.92
10,000	3.20	1.45	3.03
12,000	3.37	1.35	3.02
15,000	3.56	1.23	2.95
17,000	3.65	1.17	2.89
20,000	3.75	1.10	2.79
22,000	3.79	1.05	2.73
25,000	3.83	0.99	2.63
30,000	3.83	0.92	2.49
35,000	3.77	0.87	2.36
40,000	3.63	0.82	2.25

Table 6.
Recombination rate coefficients (cm^3/s) for ($1s^2$) ^{1}S states of H^-, He, and Li^+.

k	^1D, e-H	^3D, e-H	^1D, e-He$^+$	^3D, e-He$^+$	^1D, e-Li^{2+}	^3D, e-Li^{2+}
0.1	1.3193 (−3)	1.3217 (−3)	5.9268 (−3)	8.5133 (−3)	3.0363 (−3)	8.2703 (−3)
0.2	5.0217 (−3)	5.0835 (−3)	6.1299 (−3)	9.0331 (−3)	3.0585 (−3)	8.4642 (−3)
0.3	1.0531 (−2)	1.0898 (−2)	6.4446 (−3)	9.8834 (−3)	3.0508 (−3)	8.7011 (−3)
0.4	1.7250 (−2)	1.8401 (−2)	6.8511 (−3)	1.1044 (−2)	3.0776 (−3)	9.0700 (−2)
0.5	2.4675 (−2)	2.7204 (−2)	7.3028 (−3)	1.2473 (−2)	3.0782 (−3)	9.5041 (−2)
0.6	3.2495 (−2)	3.6934 (−2)	7.7904 (−3)	1.4152 (−2)	3.0608 (−3)	1.0009 (−2)
0.7	4.0544 (−2)	4.7286 (−2)	8.3087 (−3)	1.6066 (−2)	3.0831 (−3)	1.0622 (−2)
0.8	4.8620 (−2)	5.7990 (−2)	8.8420 (−3)	1.8172 (−2)	3.1396 (−3)	1.1380 (−2)
0.9	5.6532 (−2)	6.8791 (−2)	9.3860 (−3)	2.0439 (−2)	3.1537 (−3)	1.2151 (−2)

Table 7.
D-wave phase shifts (radians) for e-H, e-He$^+$, and e-Li^{2+}.

calculate photoionization cross sections from P-state of an atom, both S-wave and D-wave continuum functions are needed. We give a few phase shifts for ^1D and ^3D in **Table 7** for scattering from H, He$^+$, and Li^{2+}.

6. Laser fields

Scattering cross sections have also been calculated in the presence of laser fields [25]. A strong suppression in the laser-assisted cross sections is noted when compared to cross sections in the field-free situation. Further, scattering cross sections have also been calculated in the presence of Debye potential [26], in addition to the laser field.

7. Positron-hydrogen scattering

Dirac in 1928, combining the ideas of relativity and quantum mechanics, formulated the well-known relativistic wave equation and predicted an antiparticle of the electron of spin ħ/2. At that time only protons and electrons were known. He thought that the antiparticle must be proton. Hermann Weyl showed from symmetry considerations that the antiparticle must have the same mass as an electron. There are many other examples where symmetry played an important role, e.g., Newton's third law of motion (for every action there is an equal and opposite reaction) and Faraday's laws of electricity and magnetism (electric currents generate magnetic fields, and magnetic fields generate electric currents). Symmetry laws have some profound implications as shown by Emmy Noether in 1918 that every symmetry in the action is related to a conservation law [27].

Positrons, produced by cosmic rays in a cloud chamber, were detected by Anderson [28] in 1932. Positrons can form positronium atoms which annihilate, giving 511 KeV line with a width of 1.6 keV. This line has been observed from the center of the galaxy. Positrons have become very useful to scan the human brain (PET scans). They have been used to probe the Fermi surfaces, and the annihilation of positronium atoms in metals has been used to detect defects in metals.

Calculations of positron-hydrogen scattering should be simpler than the electron-hydrogen scattering because of the absence of the exchange between electrons and positrons. However, the complications arise due to the possibility of positronium atom formation. In electron-hydrogen system, the two electrons are on either side of the proton because of the repulsion between two electrons. However, because of the attraction between a positron and an electron, both the positron and the bound electron tend to be on the same side of the proton. This configuration shows that the correlations are more important in the case of a positron incident on a hydrogen atom. In 1971, we [29] carried out calculations using the projection-operator formulism of Feshbach [4] and using generalized Hylleraas-type functions:

$$\Psi(r_1, r_2, r_{12}) = e^{-\gamma r_1 - \delta r_2 - \alpha r_{12}} \sum_{lmn} C_{lmn} r_1^l r_2^m r_{12}^n. \tag{26}$$

Nonlinear parameters are γ, δ, and α and C's are the linear coefficients. The results obtained agree with those obtained by Schwartz [2]. However, the long-range correlations could not be taken into account in [29] at the same time and had to be added separately, with the result that the final phase shifts ceased to have any

k	Hybrid theory [30]	Bhatia et al. [29]	Schwarz [2]	Hybrid theory [31]	Bhatia et al. [32]	Armstrong [33]
		S-wave			*P*-wave	
0.1	0.14918	0.1483	0.151	0.008871	0.00876	0.008
0.2	0.18803	0.1877	0.188	0.032778	0.03251	0.032
0.3	0.16831	0.1677	0.168	0.06964	0.6556	0.064
0.4	0.12083	0.1201	0.120	0.10047	0.10005	0.099
0.5	0.06278	0.0624	0.062	0.13064	0.13027	0.130
0.6	0.00903	0.0039	0.007	0.15458	0.15410	0.153
0.7	0–0.04253	−0.0512	−0.54	0.17806	0.17742	0.175

Table 8.
Comparison of S*-wave and* P*-wave phase shifts obtained in the hybrid theory with results obtained earlier.*

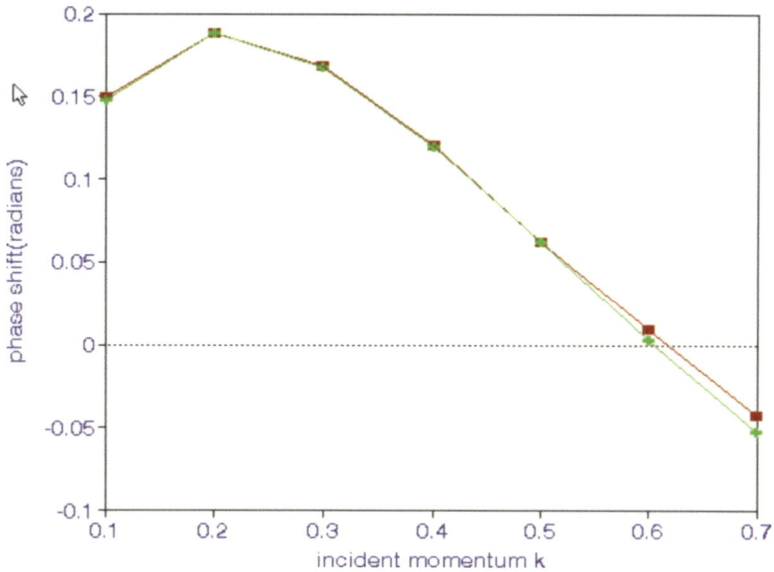

Figure 6.
The upper curve represents phase shifts obtained using the hybrid theory, and the lower curve represents phase shifts obtained using the Feshbach formalism.

bounds. In the hybrid theory, as indicated earlier, both short-range and long-range correlations can be taken into account at the same time. It should be noted that in the case of positrons, the sign before the second term in Eq. (10) is plus instead of minus, as in the case of electrons. The phase shifts obtained in two approaches are given in **Table 8**. We used 84 terms in the Hylleraas wave function previously, and improved results are obtained now with shorter expansions [30] as indicated in the table. A comparison of the results obtained with different approaches is shown in **Figure 6**. Using a fewer number of terms, higher phase shifts have been obtained in the hybrid theory [30]. P-wave shifts have been calculated in the hybrid theory [31] and are compared with those using the Feshbach projection-operator formalism [32] and with those obtained by Armstrong [33].

8. Z_{eff}

The incident positron can annihilate the atomic electron with the emission of two gamma rays. The cross section for this process has been given by Ferrell [34]:

$$\sigma_a\left(\pi a_0^2\right) = Z_{eff}\alpha^3 k^{-1} \tag{27}$$

where α is the fine-structure constant and

$$Z_{eff} = \int d\vec{r}_1 d\vec{r}_2 \left|\Psi\left(\vec{r}_1, \vec{r}_2\right)\right|^2 \delta\left(\vec{r}_1, \vec{r}_2\right). \tag{28}$$

Z_{eff} for partial waves l = 0 and 1, obtained in [31, 32], is given in **Table 9** along with the contribution from $l > 1$. For higher partial waves, Z_{eff} is given by

$$Z_{eff}(l > 1) = \frac{k^2}{1+k^2} + \frac{6}{k^2}\left[\frac{1}{k^2}\ln\left(1+k^2\right) - \frac{1+0.5k^2}{1+k^2}\right]. \tag{29}$$

k	Z_{eff} $(l = 0)$	Z_{eff} $(l = 1)$	Z_{eff} $(l > 1)$	Total
0.1	7.363	0.022	<0.001	7.385
0.2	5.538	0.90	0.001	5.629
0.3	4.184	0.187	0.004	4.375
0.4	3.327	0.294	0.010	3.631
0.5	2.730	0.390	0.022	3.142
0.6	2.279	0.464	0.039	2.782
0.7	1.850	0.528	0.063	2.541

Table 9.
Z_{eff} *for positron-hydrogen scattering.*

9. Positronium formation

Positronium, the bound state of an electron and a positron, was predicted by Mohorovicic [35] in connection with the spectra of nebulae. Positronium (Ps) formation takes place when the incident positron captures the bound electron of the hydrogen atom:

$$e^{+} + H(1s) \rightarrow Ps + P. \tag{30}$$

Cross sections for the positronium formation are given in **Table 10** and are compared with those obtained by Khan and Ghosh [36] and Humberston [37].

k^2	Hybrid theory	Khan and Ghosh [36]	Humberston [37]
0.5041	0.0066228	0.009037	0.0041
0.5476	0.018783		
0.5625	0.0.20249	0.024795	0.0044
0.64	0.022566	0.0248	0.0049
0.6724	0.022350		
0.7225	0.21456	0.021164	0.0058
0.75	0.020835	0.019707	
0.81	0.19256		
0.9025	0.016760		
1.00	0.014327		

Table 10.
Cross sections (πa_0^2) *for positronium formation obtained in hybrid theory and in comparison with those obtained by Khan and Ghosh [36] and Humberston [37].*

10. Photodetachment of positronium ion (Ps^{-})

Photodetachment has been discussed above already. Following the work of Ohmura and Ohmura [20], Bhatia and Drachman [38] calculated cross sections (in the length and velocity form) for photodetachment of Ps^{-}. Their result in length and velocity form is

$$\sigma = \left(1.32 \times 10^{-18}\text{cm}^2\right) \frac{k^3}{\left(k^2 + \gamma^2\right)^3}. \tag{31}$$

The electron affinity is $\frac{3\gamma^2}{2}$, where $\gamma = 0.12651775$ [39] and k is the momentum of the outgoing electron. These cross sections are much larger than that for the photodetachment of the negative hydrogen ion.

Lyman-α radiation ($2P \to 1S$) at 1216 Å has been seen from astrophysical sources and the sun. It has been observed from Voyager measurements [40]. Similarly, Lyman-α radiation ($2P \to 1S$) at 2416 Å is expected when in the photodetachment of Ps^-, the remaining positronium is left in the $2P$ state. Following Ohmura and Ohmura [20], photodetachment cross sections have been calculated when the remaining atom is in $2p$, $3p$, $4p$, $5p$, $6p$, and $7p$ states [41].

$$\begin{aligned} o(2p) &= 164.492C(k) \\ \sigma(3p) &= 26.3782C(k) \\ \sigma(4p) &= 9.1664C(k) \\ \sigma(5p) &= 4.3038C(k) \\ \sigma(6p) &= 2.3764C(k) \\ \sigma(7p) &= 0.2675C(k) \end{aligned} \tag{32}$$

These cross sections are given in **Figure 7** for various photon energies.

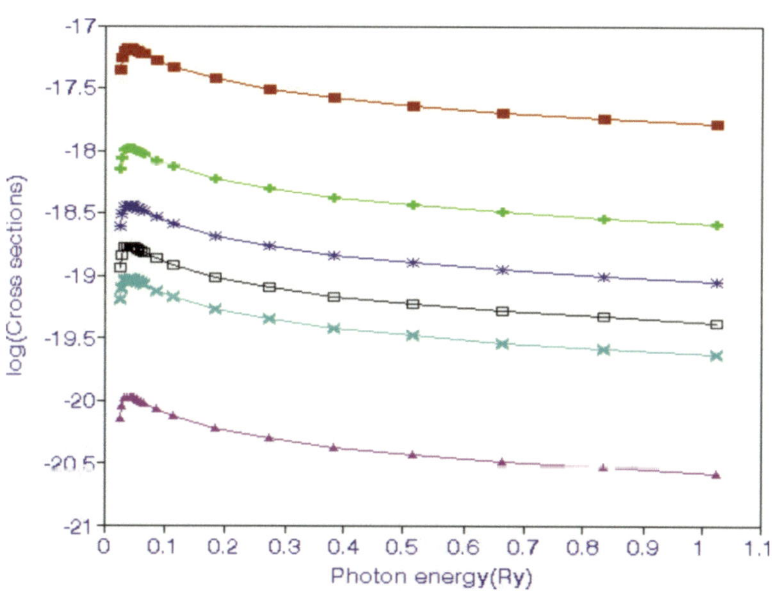

Figure 7.
The upper most curve represents photodetachment cross sections for n = 2 *on* $\log_{10}$ *scale. The curves below it are for n = 3, 4, 5, 6, and 7.*

11. Binding energies

Positrons do not bind with hydrogen atoms. However, they do bind with various atoms as has been shown by Mitroy et al. [42]. The binding energies are given in **Table 11**.

He (^3S)	Li	Be	Na	Mg
0.0011848	0.004954	0.006294	0.000946	0.031224
Ca	**Sr**	**Cu**	**Au**	**Cd**
0.03300	0.0201	0.011194	0.011664	0.01220

Table 11.
Binding energies (Ry) of positrons with various atoms.

12. Resonances

Resonances formed in the scattering of electrons from atoms are very common. However, they are not that common in positron-target systems. The first successful prediction of S-wave Feshbach resonance in positron-hydrogen system is by Doolen et al. [43], who, using the complex-rotation method, obtained the position −0.2573741 and for width 0.0000677 Ry. In this method

$$r \to re^{-i\theta}, T \to Te^{-2i\theta}, \text{and } V \to Ve^{-i\theta} \text{ and } H = T + V. \quad (33)$$

Eigenvalues are complex now. The real part gives position of the resonance, and the imaginary part gives its half width.

A number of Feshbach and shape resonances in Ps^- have been calculated by using the complex-rotation method. Parameters of a ^1P shape resonance above $n = 2$ have been calculated by Bhatia and Ho [44]. They obtained −0.06217 and 0.000225 Ry for the position and width of the resonance. These results have been confirmed experimentally by Michishio et al. [45].

13. Antihydrogen formation

Antihydrogen can be formed in the collision of Ps with antiproton

$$Ps + \overline{p} \to \overline{H} + e. \quad (34)$$

In the above equation, $\overline{p}$ represents antiproton and $\overline{H}$ represents antihydrogen. According to the time reversal invariance, the above reaction is related to

$$Ps + p \to H + e^+. \quad (35)$$

From this reaction, the positronium formation cross sections are known from positron-hydrogen scattering. This implies that the cross section for antihydrogen is related to the cross section for Ps formation:

$$\sigma_{\overline{H}} = \left(\frac{k_{e^+}}{k_{Ps}}\right)^2 \sigma_{Ps}. \quad (36)$$

k_{e^+} and k_{Ps} are momenta of positron and positronium. Humberston et al. [46] have calculated cross sections for the formation of antihydrogen in reaction (34). It is possible to form antihydrogen by radiative recombination or three-body recombination:

$$e^+ + \overline{p} \to \overline{H} + h\nu. \quad (37)$$

$$e^+ + e^+ + \overline{p} \rightarrow \overline{H} + e^+. \quad (38)$$

If the antihydrogen is formed in the excited state, then it can decay to the lower states. It would then be possible to verify if the quantum mechanics principles are the same in the antimatter universe. It is thought that gravitational interactions should be the same between matter and antimatter and between antimatter and antimatter. However, there is no experimental confirmation up to now.

14. High-energy cross sections

At high energies, only static potential remains. Therefore, according to the first Born approximation, total cross sections for e^--He and e^+-He should be the same. This fact has been verified experimentally by Kauppila et al. [47], and their results are shown in **Table 12**. We see that as the incident energy increases, cross sections tend to be equal.

Energy (eV)	e^--He	e^+-He
50	1.27	1.97
100	1.16	1.26
150	0.967	0.987
200	0.796	0.812
300	0.614	0.612
500	0.437	0.434
600	0.371	0.381

Table 12.
Cross sections measured by Kauppila et al. [47] for e^--He and e^+-He scattering.

15. Total positron-hydrogen cross sections

Total cross sections for positron scattering from hydrogen atoms have been measured by Zhou et al. [48] and have been calculated by Walters [49] using the close-coupling approximation and also by Gien [50] using the modified Glauber approximation. Their results are given in **Table 13**. They are fairly close to the experimental results.

E (eV)	Ref. [49]	Ref. [50][a]
54.4	3.02	2.85
100	2.24	2.00
200	1.33	1.24
300	9.69 (−1)	8.73 (−1)

[a]*Cross sections have been interpolated from those given in Ref. [50].*

Table 13.
Positron-hydrogen total cross sections.

16. Threshold laws

When the incident electron or positron has just enough energy to ionize the hydrogen atom, how do the cross sections behave? Wannier [51], using classical methods and supposing the two electrons emerge opposite to each other, showed that $\sigma \propto E^{1.127}$, where E is the excess energy. As indicated earlier in the positron-hydrogen scattering, positron and electron tend to be on the same side of the nucleus. This shows the cross section at threshold cannot be the same for positrons and electrons. The threshold behavior of the positron impact on hydrogen has been analyzed by Klar [52]. He finds that at threshold, $\sigma \propto E^{2.650}$.

Wigner [53] has emphasized the importance of long-range forces near the threshold which have been included in these calculations (hybrid theory). At the threshold, the cross section for exciting the 1S state of hydrogen atom to the 2S state is proportional to $(\ln k_f)^{-2}$ [54].

17. Conclusions

In this chapter, we have discussed various interactions of electrons and positrons with atoms, ions, and radiation fields. There are various approximations and theories to calculate scattering functions. Theories which provide variational bounds on the calculated phase shifts are preferable because improved results can be obtained when the number of functions in the closed channels is increased. Such theories are the close-coupling, R-matrix, and hybrid theory. It should be possible to formulate the hybrid theory for more complicated systems.

The continuum functions obtained using the hybrid theory have been used to calculate photoabsorption cross sections, obtaining results which agree with definitive results obtained using other methods and experiments. Such cross sections are needed to study the opacity in the sun. The resonances play an important role when they are included in the calculations of excitation cross sections, which are important to infer temperatures and densities of solar and astrophysical plasmas.

When Feshbach projection-operator formalism [4] is used to calculate resonance position, $E = \varepsilon + \Delta$, where Δ, the shift in the resonance position due to its coupling with the continuum, has to be calculated separately [14]. However, in the hybrid theory, the calculated position includes this correction. This is an advantage since the calculation of the shift is nontrivial.

We have indicated that in addition to obtaining accurate phase shifts for positron scattering from a hydrogen atom, we have described calculations of annihilation, positronium, and antihydrogen formation. We have discussed resonances in a positron-hydrogen system. We have discussed photodetachment of a positronium ion and a possibility of observing Lyman-α radiation from a positronium atom when the final state is the 2p state.

Author details

Anand K. Bhatia
NASA/Goddard Space Flight Center, United States

*Address all correspondence to: anand.k.bhatia@nasa.gov

IntechOpen

© 2020 The Author(s). Licensee IntechOpen. This chapter is distributed under the terms of the Creative Commons Attribution License (http://creativecommons.org/licenses/by/3.0), which permits unrestricted use, distribution, and reproduction in any medium, provided the original work is properly cited. cc BY

References

[1] Morse PM, Allis WP. The effect of exchange on the scattering of slow electrons from atoms. Physics Review. 1933;**44**:269

[2] Schwartz C. Electron scattering from hydrogen. Physics Review. 1961;**124**: 1468

[3] Temkin A. A note on scattering of electrons from atomic hydrogen. Physics Review. 1959;**116**:358

[4] Feshbach H. A unified theory of nuclear reactions II. Annals of Physics. 1962;**19**:287

[5] Burke PG, Smith K. The low-energy scattering of electrons and positrons by hydrogen atoms. Reviews of Modern Physics. 1962;**34**:458

[6] Bhatia AK. Hybrid theory of electron-hydrogen scattering. Physical Review A. 2007;**75**:032713

[7] Temkin A, Lamkin JC. Application of the method of polarized orbitals to the scattering of electrons from hydrogen. Physics Review. 1961;**121**:788

[8] Shertzer J, Temkin A. Direct calculation of the scattering amplitude without partial-wave analysis III. Inclusion of correlation effects. Physical Review A. 2006;**74**:052701

[9] Bhatia AK, Temkin A. Symmetric euler-angle decomposition of the two-electron fixed – nucleus problem. Reviews of Modern Physics. 1964;**36**: 1050

[10] Scholz T, Scott P, Burke PG. Electron-hydrogen atom scattering at intermediate energies. Journal of Physics B. 1988;**21**:L139

[11] Bhatia AK, Temkin A. Complex-correlation Kohn T-matrix method of calculating total cross sections: Electron-hydrogen elastic scattering. Physical Review A. 2001;**64**:032709

[12] Temkin A. Polarization and triplet electron-hydrogen scattering length. Physical Review Letters. 1961;**6**:354

[13] Bhatia AK. Applications of the hybrid theory to the scattering of electrons from He^+ and Li^{2+} and resonances in two-electron systems. Physical Review A. 2008;**77**:052707

[14] Bhatia AK, Temkin A. Calculation of autoionization of He and H^- using the projection-operator formalism. Physical Review A. 1975;**11**:2018

[15] Bhatia AK. Application of P-wave hybrid theory for the scattering of electrons from He^+ and resonances in He and H^-. Physical Review A. 2007;**86**: 032713

[16] Bhatia AK. Hybrid theory of P-wave electron-Li^{2+} elastic scattering and photoabsorption in two-electron systems. Physical Review A. 2013;**87**: 042705

[17] Oza DH. Phase shifts and resonances for electron scattering by He^+ below the N=2 threshold. Physical Review A. 1986; **33**:824

[18] Gien TT. Accurate calculation of phase shifts for electron collisions with positive ions. Journal of Physics B. 2003; **36**:2291

[19] Wildt R. Electron affinity in astrophysics. The Astrophysical Journal. 1939;**89**:295

[20] Ohmura T, Ohmura H. Electron-hydrogen scattering at low energies. Physics Review. 1960;**118**:154

[21] Nahar SN. In: Chavez M, Bertone E, Rosa-Gonzalez D, Rodriguez-Merino

LR, editors. New Quests Stellar Astrophysics. II. The Ultraviolet Properties of Evolved Stellar Populations. New York: Springer; 2009. p. 245

[22] Samson JAR, He ZX, Yin L, Haddad GN. Precision measurements of the absolute photoionization cross sections of He. Journal of Physics B. 1994;**27**:887

[23] Bhatia AK. Hybrid theory and its applications. Atoms. 2018;**6**:7

[24] Bhatia AK. P-wave electron-Be^{3+}, C^{5+}, and O^{7+} elastic scattering and photoabsorption in two-electron systems. Journal of Atomic, Molecular, Condensate and Nano Physics. 2014;**1**: 2349

[25] Sinha C, Bhatia AK. Laser-assisted free-free transitions in electron-atom collisions. Physical Review A. 2011;**83**: 063417

[26] Bhatia AK, Sinha C. Sinha. Free-Free transitions of the e-H system inside a dense plasma irradiated by a laser field at very low incident-electron energies. Physical Review A. 2012;**86**:053421

[27] Rodhman T, Sudarshan ECG. Doubt and Certainty, Helix Books. Mass: Reading; 1998. p. 64

[28] Anderson CD. The positive electron. Physics Review. 1933;**43**:491

[29] Bhatia AK, Temkin A, Drachman RJ, Eiserike H. Generalized hylleraas calculation of positron-hydrogen scattering. Physical Review A. 1971;**3**:1328

[30] Bhatia AK. Bhatia, positronium-hydrogen scattering, annihilation, and positronium formation. Atoms. 2016; **4**:27

[31] Bhatia AK. P-wave positronium-hydrogen scattering, annihilation, and positronium formation. Atoms. 2017;**5**:17

[32] Bhatia AK, Temkin A, Eiserike H. Rigorous precision p-wave positron-hydrogen scattering calculations. Physical Review A. 1974;**9**:219

[33] Armstrong RI. Electron-hydrogen scattering calculations. Physics Review. 1968;**171**:9

[34] Ferrell RA. Theory of positron annihilations in solids. Reviews of Modern Physics. 1956;**28**:308

[35] Mohorovicic S. Possibility of new elements and their meaning in astrophysics. Astronomische Nachrichten. 1934;**253**:93

[36] Khan A, Ghosh AS. Positronium formation in positron-hydrogen scattering. Physical Review A. 1983;**27**: 1904

[37] Humberston JW. Positronium formation in s-wave positron-hydrogen scattering. Canadian Journal of Physics. 1982;**60**:591

[38] Bhatia AK, Drachman RJ. Photodetachment of the positronium negative ion. Physical Review A. 1985; **32**:1745

[39] Bhatia AK, Drachman RJ. New Calculations of the properties of the positronium ion. Physical Review A. 1983;**28**:2523

[40] Lailement R, Quemerais E, Bertaux J-L, Sandel BR, Izamodenov V. Voyager measurements of hydrogen lyman-α diffuse emission frm milky way. Science. 2011;**334**:1665

[41] Bhatia AK. Photodetachment of the positronium negative ion with excitation in the positronium atom. Atoms. 2018;**7**:2

[42] Mitroy J, Bromley MWJ, Ryzhikh GG. Topical review: Positron and positronium binding to atoms. Journal of Physics B. 2002;**35**:R81

[43] Doolen GD, Nuttal J, Wherry CJ. Evidence of a resonance in e^{+}-H S-wave scattering. Physical Review Letters. 1978;**40**:313

[44] Bhatia AK, Ho YK. Complex-coordinate calculation of $^{1,3}P$ resonances in Ps^{-} using hylleraas functions. Physical Review A. 1990;**42**:1119

[45] Michishio K, Kanai T, Azuma T, Wade K, Mochizuki I, Hyodo T, et al. Photodetachment of positronium negative ions. Nature Communications. 2016;7:11060

[46] Humberston JW, Charlton M, Jacobson FM, Deutch BI. On antihydrogen formation in collisions of antiprotons with positronium. Journal of Physics B. 1987;**20**:347

[47] Kauppila WE, Stein TS, Smart JH, Debabneh MS, Ho YK, Dawning JP, et al. Meaurements of total scattering cross sections for intermediate-energy positrons and electrons colliding with helium, neon, and argon. Physical Review A. 1981;**24**:308

[48] Zhou S, Kappila WE, Kwan CK, Stein TS. Measurements of total cross sections for positrons and electron colliding with atomic hydrogen. Physical Review Letters. 1994;**72**:1443

[49] Walters HRJ. Positron scattering by atomic hydrogen at intermediate energies. Journal of Physics B. 1988;**21**:1893

[50] Gien TT. Total cross sections for positron-hydrogen scattering. Journal of Physics B. 1995;**28**:L321

[51] Wannier GH. The threshold law of single ionization of atoms or ions by electrons. Physical Review. 1953;**90**:817

[52] Klar H. Threshold ionization of atoms by positrons. Journal of Physics B. 1981;**14**:4165

[53] Wigner EP. On the behavior of cross sections near thresholds. Physics Review. 1948;**73**:1002

[54] Sadeghpour HR, Bohn JL, Cavagnero MJ, Esry BD, Fabrikant II, Macek JH, et al. Collisions near threshold in atomic and molecular physics. Journal of Physics B. 2000;**33**:R93

Chapter 5

Origin and Fundamentals of Perovskite Solar Cells

Mohd Quasim Khan and Khursheed Ahmad

Abstract

In the last few decades, the energy demand has been increased dramatically. Different forms of energy have utilized to fulfill the energy requirements. Solar energy has been proven an effective and highly efficient energy source which has the potential to fulfill the energy requirements in the future. Previously, various kind of solar cells have been developed. In 2013, organic–inorganic metal halide perovskite materials have emerged as a rising star in the field of photovoltaics. The methyl ammonium lead halide perovskite structures were employed as visible light sensitizer for the development of highly efficient perovskite solar cells (PSCs). In 2018, the highest power conversion efficiency of 23.7% was achieved for methyl ammonium lead halide based PSCs. This obtained highest power conversion efficiency makes them superior over other solar cells. The PSCs can be employed for practical uses, if their long term stability improved by utilizing some novel strategies. In this chapter, we have discussed the optoelectronic properties of the perovskite materials, construction of PSCs and recent advances in the electron transport layers for the fabrication of PSCs.

Keywords: methyl ammonium lead halide, perovskites light absorbers, photovoltaics, perovskite solar cells

1. Introduction

The researchers believe that the solar energy have the potential to fulfill the energy requirements [1]. The earth receive enormous amount of solar energy in the form of sunlight [2]. This sunlight can be converted to the electrical energy to fulfill our energy requirements [3]. The photovoltaic device (solar cells) can directly converted the sunlight to the electricity [4]. In previous decades, different kinds of photovoltaic devices (dye-sensitized solar cells = DSSCs, organic solar cells = OSCs, polymer solar cells, quantum-dot sensitized solar cells and perovskite solar cells) were developed [5–9]. These kinds of solar cells have attracted the scientific community due to their simple manufacturing procedure and cost-effectiveness [10].

The PSCs gained huge attention because of their excellent photovoltaic performance and low-cost [11–16]. The perovskite solar cells (PSCs) involve a perovskite light absorber layer. The perovskite is a material which has a molecular formula of ABO_3. The perovskite term was given to the calcium titanate ($CaTiO_3$). There is also another class of perovskite materials exists with molecular formula of ABX_3 (where A = Cs^+, $CH_3NH_3^+$; B = Pb^{2+} or Sn^{2+} and X = I^-, Br^- or Cl^-). This class of perovskite materials possesses excellent absorption properties, charge carrier properties and suitable band gap. In 2009, Kojima et al. [17] prepared the methyl ammonium lead halide ($MAPbX_3$; where MA = $CH_3NH_3^+$, X = halide anion) perovskite materials and investigated their optoelectronic properties. Further, authors fabricated the dye sensitized solar cells using $MAPbX_3$ visible light sensitizer [17]. The performance of the developed dye sensitized solar cells was evaluated and the device exhibited good power conversion efficiency (PCE) and open circuit voltage. The fabricated dye sensitized solar cells with $MAPbX_3$ visible light sensitizer exhibited the good PCE of ~3.8% [17]. Although, this power conversion efficiency was quite interesting but the presence of liquid electrolyte vanished this performance. Hence, it was observed that the use of alternative solid state electrolyte/hole transport material would be of great significance. In this regard, numerous strategies were developed to overcome the issue of liquid electrolyte. In this regard, a solid state electrolyte was employed by Lee et al. [18] to develop the PSCs. The developed PSCs device showed the good power conversion efficiency of 10.9%. In last few years, various novel approaches were advanced to improve the photovoltaic performance of the PSCs [19–30] and recently the best PCE of 23.3% was achieved for PSCs [31].

In this chapter, we have discussed the construction of PSCs. Recent advances in PSCs with respect to the charge collection layer/electron transport layer and future prospective have also been discussed.

2. Construction of PSCs

The fabrication of PSCs required different layers such as transparent conductive oxide coated fluorine doped tin oxide = FTO, electron transport layer (generally consists of semiconducting metal oxides), light absorber layer (perovskite), hole transport material (HTM) layer and metal contact (Au) layer. In the first step, FTO glass substrate etched with the help of zinc powder and HCl followed by the washing of the etched FTO glass substrate with acetone, DI water and 2-propanol. The compact layer of the TiO_2 deposited on to the FTO glass substrate using spin coater and annealed at ~500°C for 30–40 min. Further electron transport layer also deposited on the electrode using spin coater and annealed at ~500°C for 30–40 min. The perovskite light absorber layer deposited using spin coater and annealed at ~80–120°C for 20–60 min. Further, hole transport material (HTM) also deposited using spin coater. Finally the metal contact layer (Au) deposited using thermal evaporation approach. The construction of PSCs has been presented in **Figure 1**.

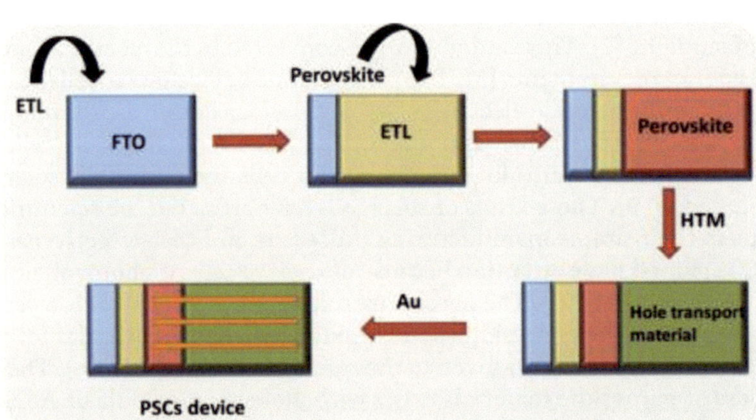

Figure 1.
Fabrication process for the PSCs.

The photovoltaic performance of the constructed PSCs can be determined by different techniques like external quantum efficiency (EQE), photocurrent-voltage (I-V), photoluminescence spectroscopy and incident-photo-to-current-conversion efficiency = IPCE etc. In general, the performance of any solar cell device can be determined in terms of fill factor, PCE, photocurrent density and open circuit voltage. Photoluminescence spectroscopy can also be employed to check the lifetime of the generated electrons inside the perovskite materials which is related to the recombination processes. It has also been known that the PSCs devices with high electron lifetime and lower recombination reaction rates may provide better photovoltaic performance in terms of power conversion efficiency.

The PSCs device absorbs the sunlight which created the electron–hole pairs inside the perovskite light absorber layer. This electron has to be transferred to the conductive electrode surface. Thus, electron transport layer consists of transition metal oxides (generally TiO_2 or SnO_2) transport this generated electron to the conductive electrode (FTO). The hole can be collected by the hole transport material layers. In some cases, the transferred electron can recombine and influence the performance of the PSCs device. Thus, some buffer or compact layers have been used to reduce the recombination process. The perovskite materials used in such PSCs devices worked as light absorber.

3. Origin of PSCs

The PSCs were originated in 2009 by Kojima et al. [17] and reported an interesting PCE of 3.1%. Further different approaches were utilized to improve the PCE of the PSCs. Recently, Tang et al. [32] prepared the low temperature processed zinc oxide nanowalls (ZnO NWs). Authors employed these prepared ZnO NWs as electron collection layer for the development of PSCs [32]. The morphological features of the prepared ZnO NWs were determined by scanning electron microscopy = SEM and transmission electron microscopy = TEM. The SEM and TEM results confirmed the formation of ZnO NWs. Further PSCs were constructed using ZnO NWs as electron collection layer whereas $MAPbI_3$ as light absorber. The device architecture of the PSCs has been presented in **Figure 2a**. The energy level values of the perovskite light absorber, ZnO, indium doped tin oxide (ITO), spiro-OMeTAD and Ag have been displayed in **Figure 2b**.

The performance of the PSCs devices with ZnO NWs and ZnO thin films were evaluated by J-V analysis. The J-V curves of the PSCs developed with ZnO NWs and ZnO thin films have been depicted in **Figure 2c**. The constructed PSCs device with ZnO NWs exhibited the highest PCE of 13.6% whereas the PSCs developed with ZnO thin films showed the PCE of 11.3%. This showed that ZnO NWs plays crucial role in charge collection compare to the ZnO thin films. The NWs of ZnO collect the generated electron more efficiently compare to the ZnO thin films. Further, incident IPCE of the constructed PSCs was also checked. The IPCE curves of the PSCs developed with ZnO NWs and ZnO thin films have been presented in **Figure 2d**. The PSCs developed with ZnO NWs showed the highest open circuit voltage of 1000 mV whereas the PSCs device fabricated with ZnO thin films showed the open circuit voltage of 980 mV. The constructed PSCs with ZnO NWs exhibited the improved IPCE compared to the PSCs device developed with ZnO thin films.

In other work, Mahmud et al. [33] synthesized low-temperature processed ZnO thin film.

The optical properties of the prepared ZnO thin film were investigated by employing ultraviolet–visible (UV–vis) absorption spectroscopy. The Tauc plot of the ZnO has been presented in **Figure 3A**. The synthesized ZnO possess an optical

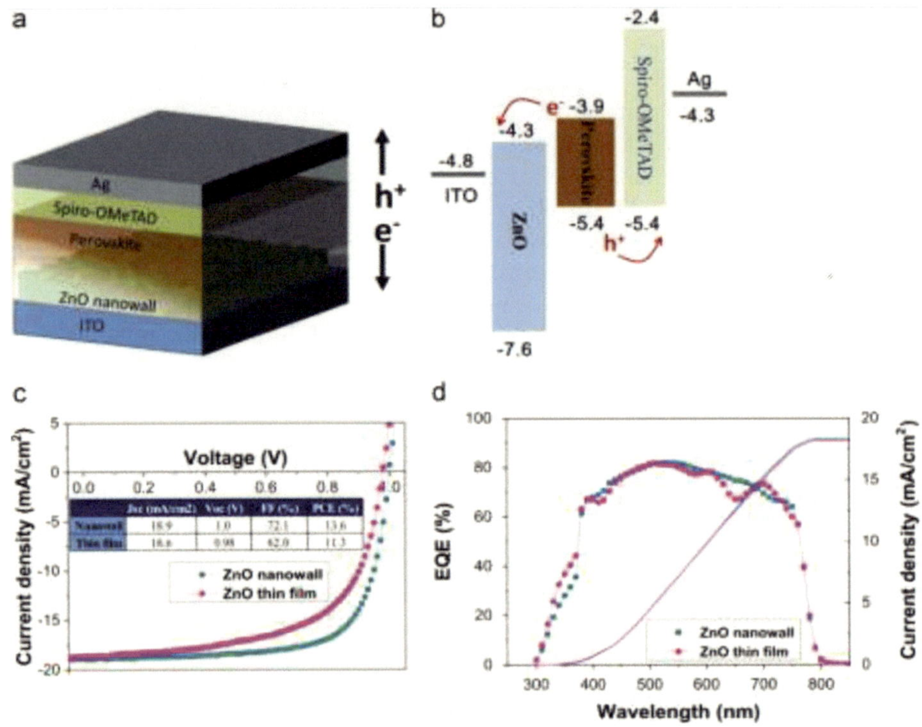

Figure 2.
Schematic picture of PSCs device architecture (a). Energy level diagram of PSCs components (b). J-V graphs of the PSCs constructed with ZnO NWs and ZnO thin films (c). IPCE of the PSCs constructed with ZnO NWs and ZnO thin films (d). Reprinted with permission [32].

band gap of 3.53 eV as confirmed by Tauc relation. The formation of ZnO on ITO glass substrate was confirmed by employing X-ray diffraction = XRD method. The XRD pattern of the ZnO has been presented in **Figure 3B**. The XRD pattern of the ZnO showed the crystalline nature with strong diffraction peaks. Authors employed ZnO thin film as electro transport layer for the construction of PSCs [33]. The $MAPbI_3$ was utilized as light absorber layer. Authors also investigated the morphological features of the $MAPbI_3$ films prepared on ZnO. The SEM results showed the presence of uniform surface morphology of the $MAPbI_3$ perovskite [33]. Further, PSCs were fabricated and the device architecture of the fabricated PSCs has been depicted in **Figure 4A**.

The energy level diagram of the fabricated PSCs device has been presented in **Figure 4B**. The photovoltaic performance of the constructed PSCs device was evaluated by recording J-V curve. The obtained results showed that the fabricated PSCs device with ZnO thin film possess a highest PCE of 8.77% with open circuit voltage of 932 mV.

In 2017, Li et al. [34] synthesized ZnO/Zn_2SnO_4 under facile conditions. The synthesized ZnO/Zn_2SnO_4 was utilized as compact layer for the fabrication of $MAPbI_3$ based PSCs. Authors recorded the XRD pattern of the $MAPbI_3$ perovskite layer [34]. The XRD pattern of the $MAPbI_3$ perovskite layer has been presented in **Figure 5a**.

The XRD pattern of the $MAPbI_3$ perovskite layer showed the well-defined diffraction planes which suggested the successful formation of $MAPbI_3$ perovskite as shown in **Figure 5a**. The formation of the ZnO/Zn_2SnO_4 was checked by XRD and X-ray photoelectron spectroscopy (XPS). The recorded XRD pattern of the ZnO/Zn_2SnO_4 has been presented in **Figure 5b**. The XRD pattern showed the diffraction planes for the ZnO, Zn_2SnO_4 and SnO_2

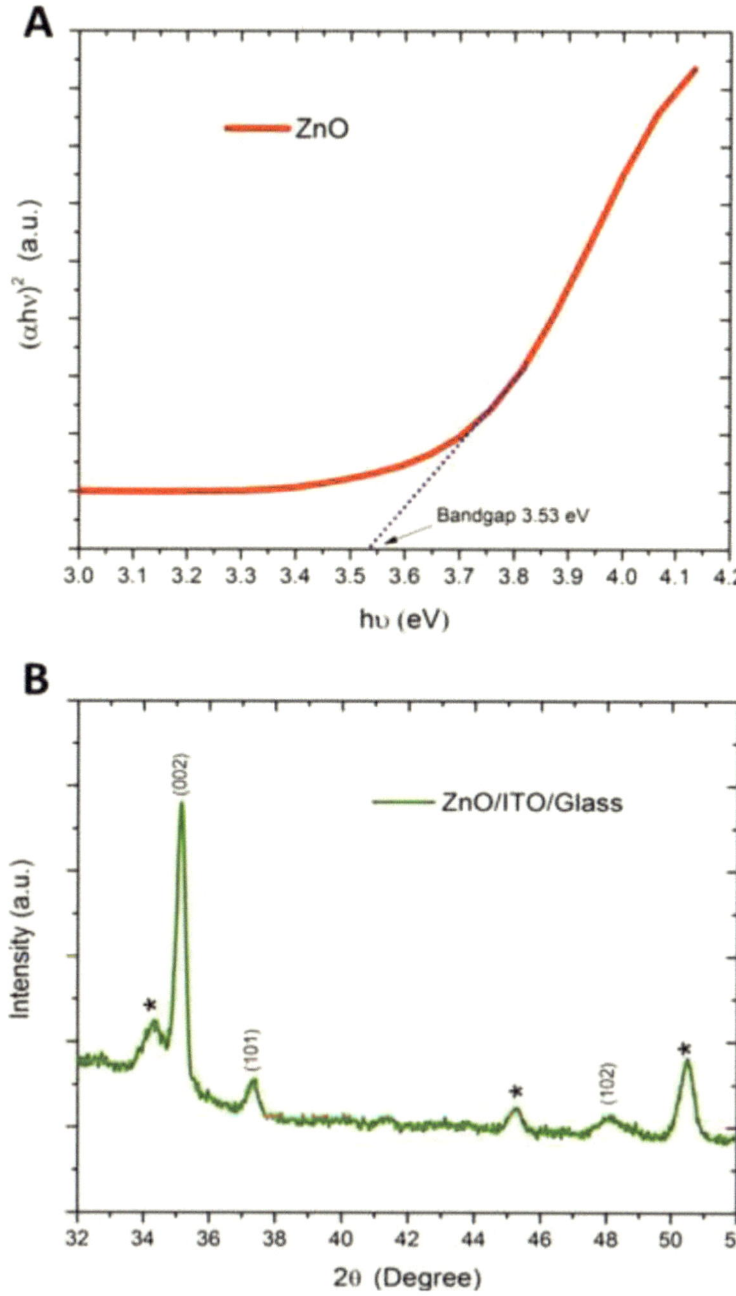

Figure 3.
Tauc plot of the ZnO (A). XRD pattern of the ZnO/ITO (B). Reprinted with permission [33].

This confirmed the formation of ZnO/Zn_2SnO_4. Further, authors also investigated the morphological characteristics of the ZnO/Zn_2SnO_4 using SEM analysis [34]. Authors employed ZnO/Zn_2SnO_4 as compact layer and developed the PSCs devices [34]. Authors also developed the PSCs using TiO_2 with different thickness [34]. The performance of the developed PSCs devices were evaluated by J-V approach. The recorded J-V curves of the developed PSCs with different thicknesses (40 nm, 60 nm, 80 nm, 100 nm and 120 nm) of TiO_2 have been presented in **Figure 6**. The PSCs device fabricated with TiO_2 (thickness = 100 nm) exhibited the highest performance compared to the PSCs device fabricated with TiO_2 of different thicknesses.

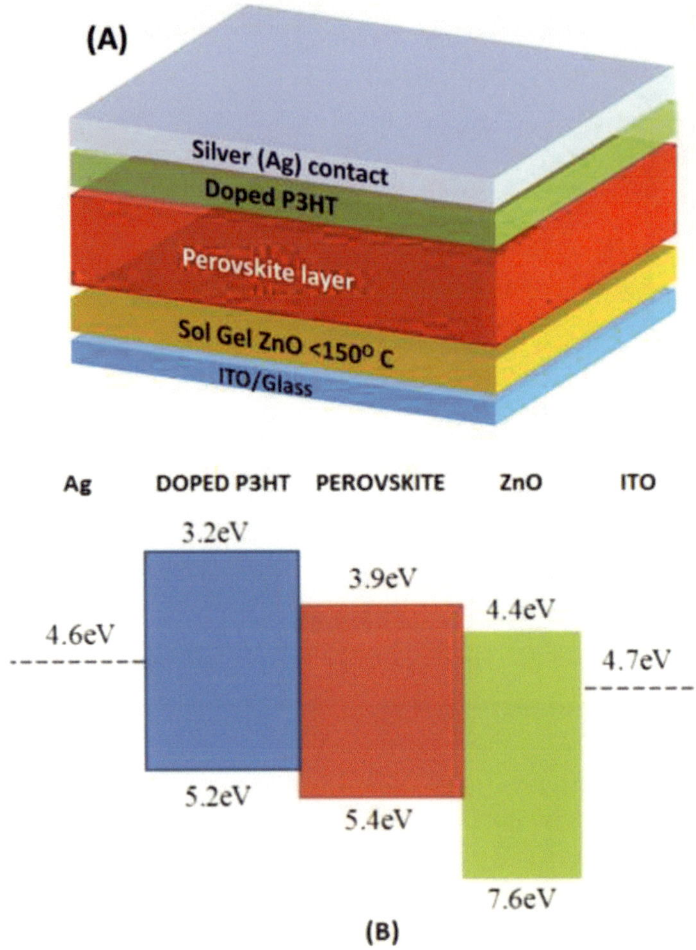

Figure 4.
Schematic diagram of the PSCs device (A). Energy level diagram of the PSCs (B). Reprinted with permission [33].

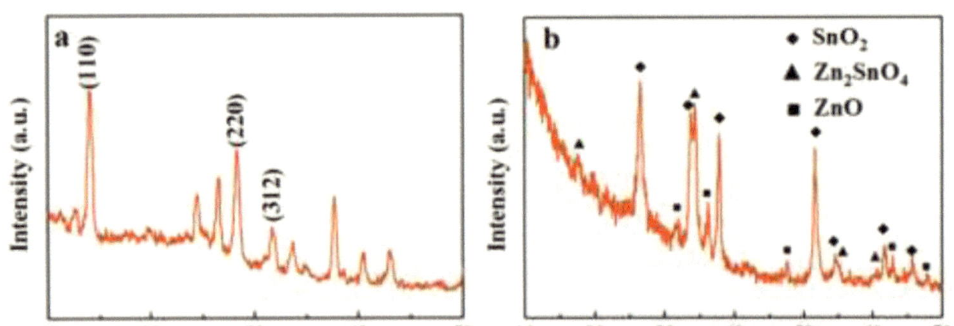

Figure 5.
XRD patterns of the MAPbI3 (a) and ZnO/ZSO CL (b). Reprinted with permission [34].

Furthermore, the photovoltaic performance of the PSCs developed using ZnO/ Zn_2SnO_4 as compact layer with different thickness (15 nm, 35 nm, 55 nm, 75 nm and 95 nm) were also evaluated. The J-V curves of the PSCs developed using ZnO/

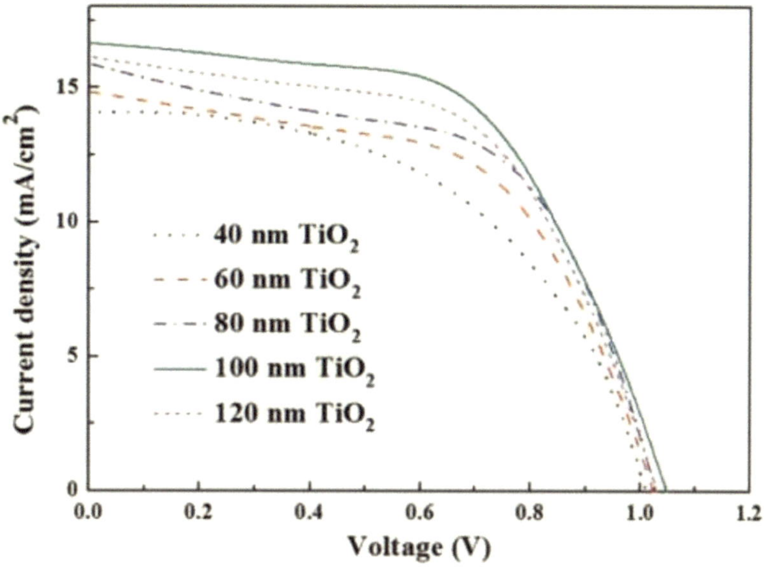

Figure 6.
J-V curves of PSCs based on different thickness of TiO_2 CLs. Reprinted with permission [34].

Zn_2SnO_4 with different thickness (15 nm, 35 nm, 55 nm, 75 nm and 95 nm) has been presented in **Figure 7**. Authors found that the PSCs device fabricated with ZnO/ Zn_2SnO_4 (thickness = 15 nm) has poor photovoltaic parameters which resulted to the poor performance [34].

The PSCs device fabricated with ZnO/Zn_2SnO_4 (thickness = 75 nm) showed enhanced photovoltaic parameters which resulted to the improved photovoltaic performance (**Figure 7**). This showed that ZnO/Zn_2SnO_4 (thickness = 75 nm) is more effective charge compact layer compared to the TiO_2.

In another recent work, Chang et al. [35] developed the PSCs using Ce doped $CH_3NH_3PbI_3$ perovskite light absorber layer.

In this work, Chang et al. [35] prepared the thin films of Ce doped $CH_3NH_3PbI_3$ perovskite light absorber layer using a post treatment approach. Authors used CsI to promote the morphological features and crystallization of the thin films of Ce doped $CH_3NH_3PbI_3$ perovskite light absorber layer. The use of Cs helps to obtain the large grain size of the $CH_3NH_3PbI_3$ perovskite. The grain size of the $CH_3NH_3PbI_3$ perovskite absorber layers were ranges 270 nm–650 nm. The formation of the perovskite light absorber layers were confirmed by XRD analysis. The optical band gap of the perovskite light absorber layer was also calculated by using Tauc relation.

The Cs doped $CH_3NH_3PbI_3$ perovskite light absorber has a band gap of 1.59 eV whereas this band gap slightly increases with increasing CsI concentrations. The increase in the optical band gap of the $CH_3NH_3PbI_3$ perovskite absorber layer also confirmed the insertion of Cs in to the perovskite light absorber layer.

The SEM pictures of the $CH_3NH_3PbI_3$ perovskite light absorber layers were also recorded. The recorded SEM pictures of the $CH_3NH_3PbI_3$ perovskite light absorber layers without and with CsI treatment have been presented in **Figure 8a–f**. The SEM picture of the $CH_3NH_3PbI_3$ perovskite light absorber layer without CsI treatment showed the small grain size (**Figure 8a**). However, the insertion of CsI to the $CH_3NH_3PbI_3$ perovskite light absorber layer increases the grain size as confirmed by the SEM investigations. The highly uniform surface morphology was observed in case of $CH_3NH_3PbI_3$ perovskite absorber layer treated with 6mg mL^{-1} CsI (**Figure 8d**). Furthermore, PSCs devices were fabricated using $CH_3NH_3PbI_3$ perovskite light absorber layers. The schematic picture of the developed PSCs device has been presented in **Figure 9**. The

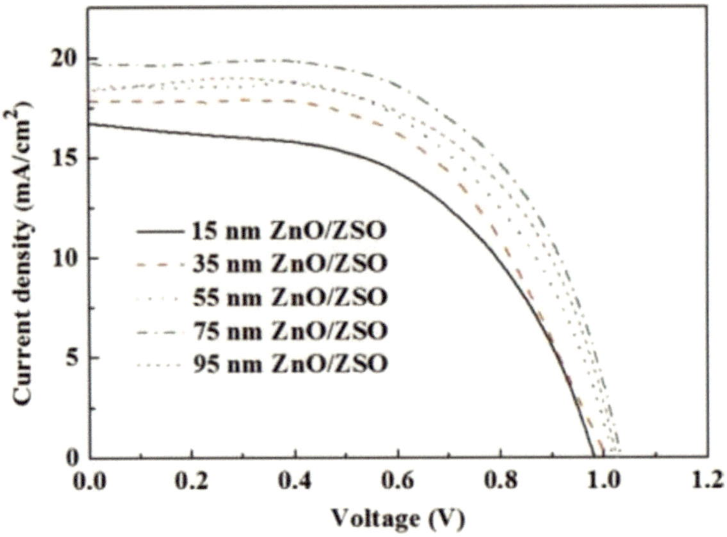

Figure 7.
J-V curves of PSCs based on different thickness of ZnO/ZSO CLs. Reprinted with permission [34].

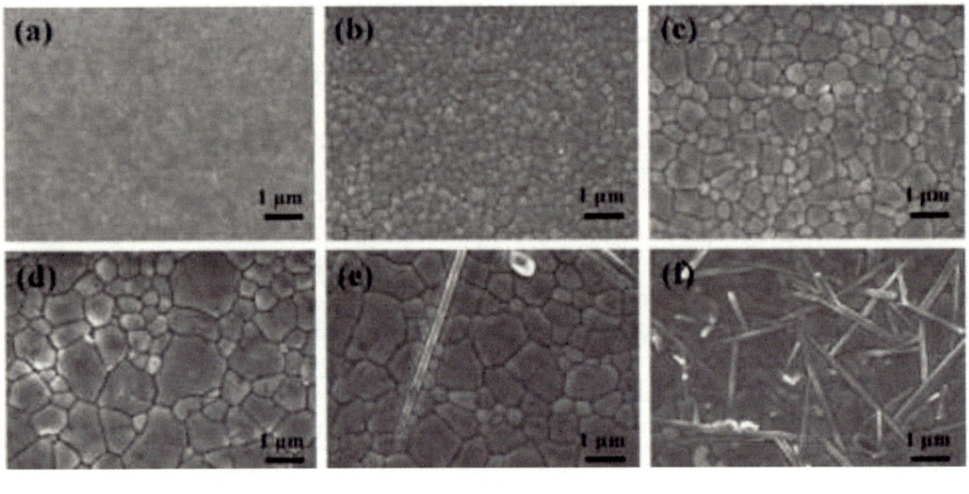

Figure 8.
SEM pictures of the $CH_3NH_3PbI_3$ thin films: untreated (a) and treated with 2.5 mg mL^{-1} (b), 5 mg mL^{-1} (c), 6 mg mL^{-1} (d), 7 mg mL^{-1} (e), 9 mg mL^{-1} CsI (f). Reprinted with permission [35].

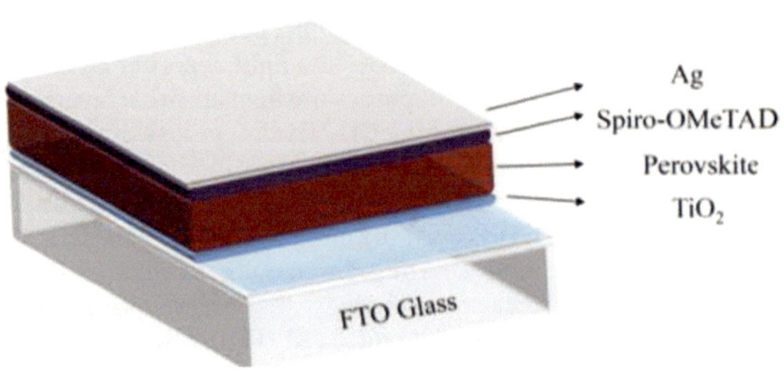

Figure 9.
Schematic picture of the constructed PSCs device. Reprinted with permission [35].

Absorber layer	J_{SC} (mA/ cm^2)	V_{OC} (mV)	PCE (%)	Type of solar cells	References
$MAPbI_3$	19.2	720	10.2	PSCs	[36]
perovskite	22.7	240	2.02	PSCs	[37]
$MASnI_3$	16.8	880	6.4	PSCs	[38]
$FASnI_3$	17.53	600	6.7	PSCs	[39]
$FASnI_3$	24.1	520	9	PSCs	[40]
$MASnI_3$	11.1	970	7.6	PSCs	[41]
$FASn_{0.5}Pb_{0.5}I_3$	21.9	700	10.2	PSCs	[42]
$MASn_{0.25}Pb_{0.75}$	15.8	730	7.37	PSCs	[43]
Al^{3+}doped $CH_3NH_3PbI_3$	22.4	1001	19.1	PSCs	[44]
Dye	13.2	570	4.63	DSSCs	[45]
Dye	15.46	821	8.20	DSSCs	[46]
Polymer light absorber	20.65	946	14.45	Polymer	[47]
Polymer light absorber	19.1	990	11.5	Polymer	[48]
Perovskite quantum dot	15.1	1220	13.8	Quantum dot PSCs	[49]
Quantum dot light absorber	26.70	780	13.84	Quantum dot solar cells	[50]
Organic light absorbing material	21.8	940	14.8	Organic solar cells	[51]

Table 1.
Comparison of the photovoltaic parameters of the PSCs with other reported solar cells.

constructed PSCs device with $CH_3NH_3PbI_3$ perovskite absorber layer (with 6 mg mL^{-1} CsI treatment) exhibited the best PCE of 14.4% with open circuit voltage of 1.05 V. However, the PSCs developed without CsI treatment showed the relatively lower PCE of 10.5%. There are different kinds of solar cells existed and each type of solar cell has different light absorbing materials. The photovoltaic performance of the PSCs has been compared with other reported solar cells as shown in **Table 1**.

4. Future prospective

Since the origin, PSCs have received enormous attention due to their simple solution-processed fabrication, high performance and cost-effectiveness. This is because of the excellent optoelectronic properties of the organic–inorganic lead halide perovskite light absorbers. The PSCs achieved a highest power conversion efficiency of more than 24%. The PSCs can be employed for practical applications due to their high performance and cost-effectiveness. However, the poor aerobic stability and moisture sensitivity of the perovskite light absorbers restricts their practical applications. Thus, it is of great importance to overcome the issue of moisture sensitivity and poor stability of the PSCs. In previous years numerous strategies and methods were developed to enhance the stability of the PSCs. However, further improvements are necessary to commercialize the PSCs at large scale.

We believe that the following points/strategies would be beneficial to further enhance the stability and photovoltaic performance of the PSCs:

1. New device architectures are required to develop the highly efficient PSCs.

2. The photovoltaic parameters/performance of the PSCs can be further improved by utilizing/developing new electron transport/charge extraction layers.

3. Some novel hydrophobic cationic groups should be introduced to the perovskite light absorbers to improve the aerobic stability and moisture sensitivity.

5. Conclusions

In present scenario, energy crisis is the major challenge for today's world. Solar cells have the potential to overcome the issue of energy crisis. In last 10 years, PSCs have attracted the materials scientists due to its excellent photovoltaic performance and easy fabrication procedures. The highly efficient PSCs involve $MAPbX_3$ as light absorber layer. The photovoltaic performance of the PSCs can be influenced by the presence of absorber layer or electron transport layer. Previously different kinds of electron transport layers have been widely studied to enhance the performance of the $MAPbX_3$ based PSCs. The highest PCE of more than 24% has been certified by NREL for $MAPbX_3$ based PSCs device. This excellent PCE is close to the commercialized silicon based solar cells. Thus, it can be said that PSCs can fulfill our energy requirements in the future. In this chapter, the fabrication of PSCs has been discussed. The recent advances in the development of PSCs with different compact layers, electron transport layers and charge collection layers have been reviewed.

Acknowledgements

K.A. would like to acknowledge Discipline of Chemistry, IIT Indore. M.Q.K. acknowledged Department of Chemistry, Faculty of Applied Science and Humanities, Invertis University.

Conflict of interest

"The authors declare no conflict of interest."

Author details

Mohd Quasim Khan[1] and Khursheed Ahmad[2*]

1 Department of Chemistry, Faculty of Applied Science and Humanities, Invertis University, Bareilly, 243123, U.P., India

2 Discipline of Chemistry, Indian Institute of Technology Indore, Simrol, Khandwa Road, 453552, M.P., India

*Address all correspondence to: khursheed.energy@gmail.com

IntechOpen

© 2020 The Author(s). Licensee IntechOpen. This chapter is distributed under the terms of the Creative Commons Attribution License (http://creativecommons.org/licenses/by/3.0), which permits unrestricted use, distribution, and reproduction in any medium, provided the original work is properly cited.

References

[1] Reddy VS, Kaushik SC, Ranjan KR, Tyagi SK. State-of-the-art of solar thermal power plants. Renewable and Sustainable Energy Reviews. 2019;**27**:258-273

[2] Chen GY, Seo J, Yang CH, Prasad PN. Nanochemistry and nanomaterials for photovoltaics. Chemical Society Reviews. 2013;**42**:8304-8338

[3] Motlak M, Hamza AM, Hammed MG, Barakat NAM. Cd-doped TiO_2 nanofibers as effective working electrode for the dye sensitized solar cells. Materials Letters. 2019;**246**:206-209

[4] O'Regan B, Grätzel M. A low cost, high-efficiency solar cell based on dye-sensitized colloidal TiO_2 films. Nature. 1991;**353**:737-740

[5] Colombo A, Dragonetti C, Roberto D, Ugo R, Manfredi N, Manca P, et al. A carbon doped anatase TiO_2 as a promising semiconducting layer in Ru-dyes based dye-sensitized solar cells. Inorganica Chimica Acta. 2019;**489**:263-268

[6] Zhang X, Liu F, Huang QL, Zhou G, Wang Z-S. Dye-sensitized W-doped TiO_2 solar cells with a tunable conduction band and suppressed charge recombination. Journal of Physical Chemistry C. 2011;**115**:12665-12671

[7] Lu WH, Chou C-S, Chen C-Y, Wu P. Preparation of Zr-doped mesoporous TiO_2 particles and their applications in the novel working electrode of a dye-sensitized solar cell. Advanced Powder Technology. 2017;**28**:2186-2197

[8] Xiang P, Lv F, Xiao T, Jiang L, Tan X, Shu T. Improved performance of quasi-solid-state dye-sensitized solar cells based on iodine-doped TiO_2 spheres photoanodes. J. Alloy Compound. 2018;**741**:1142-1147

[9] Tran VA, Truong TT, Phan TAP, Nguyen TN, Huynh TV, Agrestic A, et al. Application of nitrogen-doped TiO_2 nano-tubes in dye-sensitized solar cells. Applied Surface Science. 2017;**399**:515-522

[10] Deng J, Wang M, Fang J, Song X, Yang Z, Yuan Z. Synthesis of Zn-doped TiO_2 nano-particles using metal Ti and Zn as raw materials and application in quantum dot sensitized solar cells. J. Alloy Compound. 2019;**791**:371-379

[11] Ahmad K, Ansari SN, Natarajan K, Mobin SM. A two-step modified deposition method based $(CH_3NH_3)_3Bi_2I_9$ perovskite: Lead free, highly stable and enhanced photovoltaic performance. ChemElectroChem. 2019;**6**:1192-1198

[12] Ahmad K, Ansari SN, Natarajan K, Mobin SM. Design and synthesis of 1D-polymeric chain based $[(CH_3NH_3)_3Bi_2Cl_9]_n$ perovskite: A new light absorber material for lead free perovskite solar cells. ACS Applied Energy Materials. 2018;**01**:2405-2409

[13] Ahmad K, Mobin SM. Graphene oxide based planar heterojunction perovskite solar cell under ambient condition. New Journal of Chemistry. 2017;**41**:14253-14258

[14] Ahmad K, Mohammad A, Mobin SM. Hydrothermally grown α-MnO_2 nanorods as highly efficient low cost counter-electrode material for dye-sensitized solar cells and electrochemical sensing applications. Electrochimica Acta. 2017;**252**:549-557

[15] Zhong M, Liang Y, Zhang J, Wei Z, Li Q, Xu D. Highly efficient flexible $MAPbI_3$ solar cells with a fullerene derivative-modified SnO_2layer as the electron transport layer. Journal of Materials Chemistry A. 2019;**7**:6659-6664

[16] Guo Z, Liguo G, Zhang C, Xu Z, Ma T. Low-temperature processed non-TiO2 electron selective layers for perovskite solar cells. Journal of Materials Chemistry A. 2018;**6**:4572-4589

[17] Kojima A, Teshima K, Shirai Y, Miyasaka T. Organometal halide perovskites as visible-light sensitizers for photovoltaic cells. Journal of the American Chemical Society. 2009;**131**:6050-6051

[18] Lee MM, Teuscher J, Miyasaka T, Murakami TN, Snaith HJ. Efficient hybrid solar cells based on meso-superstructured organometal halide perovskites. Science. 2012;**338**:643-647

[19] Chen Y-Z, Wu R-J, Lin LY, Chang WC. Novel synthesis of popcorn-like TiO_2 light scatterers using a facile solution method for efficient dye-sensitized solar cells. Journal of Power Sources. 2019;**413**:384-390

[20] Im JH, Lee CR, Lee JW, Park SW, Park NG. 6.5% efficient perovskite quantum-dot-sensitized solar cell. Nanoscale. 2011;**3**:4088-4093

[21] Kim HS, Lee CR, Im JH, Lee KB, Moehl T, Marchioro A, et al. Lead iodide perovskite sensitized all-solid-state submicron thin film mesoscopic solar cell with efficiency exceeding 9%. Scientific Reports. 2012;**2**:591

[22] Wehrenfennig C, Liu M, Snaith HJ, Johnston MB, Herz LM. Charge-carrier dynamics in vapour-deposited films of the organolead halide perovskite CH3NH3PbI3−xClx. Energy & Environmental Science. 2014;**7**:2269-2275

[23] Ma J, Guo X, Zhou L, Lin Z, Zhang C, Yang Z, et al. Enhanced planar perovskite solar cell performance via contact passivation of TiO2/perovskite interface with NaCl doping approach. ACS Appl. Energy Mater. 2018;**1**:3826-3834

[24] Ke W, Fang G, Wang J, Qin P, Tao H, Lei H, et al. Perovskite solar cell with an efficient TiO2 compact film. ACS Applied Materials & Interfaces. 2014;**6**:15959-15965

[25] Peng G, Wu J, Wu S, Xu X, Ellis JE, Xu G, et al. Perovskite solar cells based on bottom-fused TiO2 nanocones. Journal of Materials Chemistry A. 2016;**4**:1520-1530

[26] Lv M, Lv W, Fang X, Sun P, Lin B, Zhang S, et al. Performance enhancement of perovskite solar cells with a modified TiO2 electron transport layer using Zn-based additives. RSC Advances. 2016;**6**:35044-35050

[27] Liu D, Kelly TL. Perovskite solar cells with a planar heterojunction structure prepared using room-temperature solution processing techniques. Nat. Photon. 2014;**8**:133-138

[28] Jeong S, Seo S, Park H, Shin H. Atomic layer deposition of a SnO2 electron-transporting layer for planar perovskite solar cells with a power conversion efficiency of 18.3%. Chemical Communications. 2019;**55**:2433-2436

[29] Wang S, Zhu Y, Liu B, Wang C, Ma R. Introduction of carbon nanodots into SnO2 electron transport layer for efficient and UV stable planar perovskite solar cells. Journal of Materials Chemistry. 2019;**7**:5353-5362

[30] Ding B, Huang SY, Chu QQ, Li Y, Li CX, Li CJ, et al. Low-temperature SnO2-modified TiO2 yields record efficiency for normal planar perovskite solar modules. Journal of Materials Chemistry A. 2018;**6**:10233-10242

[31] Zimmermann I, Aghazad S, Nazeeruddin MK. Angewandte Chemie, International Edition. 2018;**57**:1-6

[32] Tang J-F, Tseng Z-L, Chen L-C, Chu S-Y. ZnO nanowalls grown at low-temperature for electron collection

in high-efficiency perovskite solar cells. Sol. Energ. Mater. Sol. Cells. 2016;**154**:18-22

[33] Mahmud MA, Kumar N, et al. Low temperature processed ZnO thin film as electron transport layer for efficient perovskite solar cells. Sol. Energ. Mater. Sol. Cells. 2017;**159**:251-264

[34] Li W, Jiang Q, Yang J, Luo Y, Li X, Hou Y, et al. Improvement of photovoltaic performance of perovskite solar cells with a ZnO/Zn_2SnO_4 composite compact layer. Sol. Energ. Mater. Sol. Cells. 2017;**159**:143-150

[35] Chang R, Yan Y, Zhang J, Zhu Z, Gu J. Large-grain and smooth cesium doped $CH_3NH_3PbI_3$ perovskite films by cesium iodide post-treatment for improved solar cells. Thin Solid Films. 2020;**712**:138279

[36] Chung I, Lee B, He J, Chang RPH, Kanatzidis MG. All-solid-state dye-sensitized solar cells with high efficiency. Nature. 2012;**485**:486-489

[37] Kumar MH, Dharani S, Leong WL, Boix PP, Prabhakar RR, Baikie T, et al. Lead-free halide perovskite solar cells with high photocurrents realized through vacancy modulation. Advanced Materials. 2014;**26**:7122-7127

[38] Noel NK, Stranks SD, Abate A, Wehrenfennig C, Guarnera S, Haghighirad A, et al. Environmental Science for photovoltaic applications. Energy Environ Sci. 2014;**7**:3061-3068

[39] Tai Q, Guo X, Tang G, You P, Ng TW, Shen D, et al. Antioxidant grain passivation for air-stable tin-based perovskite solar cells. Angew. Chem. Int. Ed. 2019;**58**:806-810

[40] Shao S, Liu J, Portale G, Fang HH, Blake GR, Brink GH, et al. Highly reproducible Sn-based hybrid perovskite solar cells with 9% efficiency. Advanced Energy Materials. 2018;**8**:1702019

[41] Bansode U, Naphade R, Game O, Agarkar S, Ogale S. Hybrid perovskite films by a new variant of pulsed excimer laser deposition: a roomerature dry process. Journal of Physical Chemistry C. 2015;**119**:9177-9185

[42] Eperon GE, Eperon GE, Leijtens T, Bush KA, Prasanna R, Green T, et al. 2016. Eperon. 2016;**9717**:1-10

[43] Hao F, Stoumpos K, Cao DH, Chang RPH, Kanatzidis M. Lead-free solid state organic-inorganic halide perovskite solar cells. Nature Photonics. 2014;**8**:489-494

[44] Ramirez I, Zhang J, Ducati C, Grovenor C, Johnston MB, Ginger DS, et al. Environmental science. Energy & Environmental Science. 2016;**9**:2892-2901

[45] Lee S, Kang D. Highly efficient and stable sn-rich perovskite solar cells by introducing bromine. ACS Applied Materials & Interfaces. 2017;**9**:22432-22439

[46] Jia H-L, Li S-S, Gong B-Q, Gu L, Bao Z-L, Guan M-Y. Efficient co sensitization of new organic dyes containing bipyridine anchors with porphyrins for dye-sensitized solar cells. Sustainable Energy & Fuels. 2020;**4**:347-353

[47] Fan Q, An Q, Lin Y, Xia Y, Li Q, Zhang M, et al. Over 14% efficiency all-polymer solar cells enabled by a low bandgap polymer acceptor with low energy loss and efficient charge separation. Energy & Environmental Science. 2020. DOI: 10.1039/D0EE01828G

[48] Dong X, Guo Q, Liu Q, Zhu L, Guo X, Liu F, et al. A naphthodithiophene-based nonfullerene acceptor for high-performance polymer solar cells with a small energy loss. Journal of Materials Chemistry C. 2020;**8**:6513-6520

[49] Ji K, Yuan J, Li F, Shi Y, Ling X, Zhang X, et al. High-efficiency perovskite quantum dot solar cells benefiting from a conjugated polymer-quantum dot bulk heterojunction connecting layer. Journal of Materials Chemistry A. 2020;**8**:8104-8112

[50] Rao H, Zhou M, Pan Z, Zhong X. Quantum dot material engineering boosting quantum dot sensitized solar cells efficiency over 13%. Journal of Materials Chemistry A. 2020;**8**:10233-10241

[51] Wen S, Li Y, Zheng N, Raji IO, Yang C, Bao X. High-efficiency organic solar cells enabled by halogenation of polymers based on 2D conjugated benzobis(thiazole). Journal of Materials Chemistry A. 2020;**8**:13671-13678

Section 3

Semiconductor Devices

Chapter 6

Diffusion and Quantum Well Intermixing

Thamer Tabbakh

Abstract

Diffusion or intermixing is the movement of particles through space. It primarily occurs in every form of matter because of thermal motion. Atom diffusion and intermixing can also happen in crystalline semiconductors whereby the atoms that are diffusing and intermixing move from one side of the lattice to the adjacent one in the crystal semiconductor. Atom diffusion, which may also involve defects (including native and dopant), is at the core of processing of semiconductors. The stages involved in semiconductor processing are growth, followed by post-growth, and then the construction stage comes last. The control of every aspect of diffusion is necessary to accomplish the required goals, therefore creating a need for knowing what diffuses at any point in time. This chapter will briefly summarize the techniques that are in existence and are used to create diffused quantum wells (QWs). Also, it will outline the examples of QW semiconductor lasers and light-emitting diode (LED) by the utilization of inter-diffusion techniques and give recent examples.

Keywords: intermixing, semiconductors, diffusion, QWI, lasers, LED, intermixing techniques, inter-diffusion, fabrication

1. Introduction

The demands of device technology have yielded the primary motivation for looking over atomic diffusion, as depicted by a semiconductor lattice. Since there has been a shrinking of the devices' physical dimensions, more problems have emerged concerning comprehending features of diffusion in more complex structures [1]. There is a link between some common problems with the deterioration of a doped structure, for instance, a superlattice or p-n junction, diffusion barrier, or a metal contact's endurance [1–4].

Knez pointed out four diffusion situations that are separate from each other, which can crop up in the post-processing of the substrate's surface layer. The layer can be thin like mercury telluride (HgTe) or cadmium telluride (CdTe) [4–6]. There are four different diffusion situations for the post-processing, which are the following:

Firstly, there is components' lateral diffusion in the surface layer.

Secondly, there is surface component diffusion into the substrate (surface into substrate).

Thirdly, there is substrates' component diffusion into the surface layer (substrate to surface).

IntechOpen

Fourthly is the diffusion barrier stationed between the substrate and surface layer.

The type of diffusion within the crystal lattice is called lattice diffusion, and it takes place by either substitutional or interstitial mechanisms. Interstitial lattice diffusion involves a diffusant like carbon in an iron combination diffusing in the middle of the lattice structure of one or more crystalline elements. On the other hand, substitutional lattice diffusion involves self-diffusion or inter-diffusion (where self-diffusion takes place in pure metals because atoms exchange location for the same type and there is no net mass transport, while inter-diffusion occurred in alloys which have net mass transport and atoms diffuse into different metals) whereby the movement of an atom is made possible by its substitution with another atom to replace it [6–10]. This diffusion is usually made possible by point vacancies' availability all over the crystal lattice. Diffusing particles relocate fast from one vacancy point to another, basically by random jumping termed as jump diffusion, as shown in **Figure 1**. Considering that the regularity of point vacancies multiplies in line with the Arrhenius equation, the frequency of diffusion crystal solid state improves with temperature [11–15].

The use of inter-diffusion of quantum wells (QWs) is an emerging technology that is significant for fabricating semiconductor lasers since it improves devices' optical and electrical properties [16]. Selective inter-diffusion is achievable by obscuring into the QW wafer's desired regions. Since the 1980s, there have been extensive investigations regarding inter-diffusion [16, 17]. It comprises disordering or intermixing of heterostructures that are quantum-confined like QWs and quantum dots (QDs). The thorough investigations are due to its potential to achieve monolithic integration of optoelectronic/photonic devices. Among the inter-diffusion techniques, there has been a consideration of impurity-free vacancy disordering (IFVD) as the technique that is most promising for device applications because of its simplicity and causes lesser residual damage to the sample [17–19].

During inter-diffusion, there will be a resultant modification of refractive index and electrical conductivity between the regions that are as-grown together with disordered ones. The technology allows a homogenous process that leads to the enhancement of the sideways electrical and optical restraint of laser semiconductors in such a manner that the bottom threshold current, as well as single operation that is lateral mode, is obtainable. Moreover, the QW's shape alters as a result of inter-diffusion between QWs and barriers that are next to it. In turn, there is a modification of the sub-band energy in valence and conduction bands. Eventually, the inter-band transition energy is modified. Therefore, the inter-diffusion technique could be utilize the fabrication of QW lasers and LED for multiple wavelengths without using complicated epitaxial regrowth or etching processes. Other merits of utilizing inter-diffusion techniques include one, its simplicity. And there is also compatibility with existing semiconductor lasers' fabrication technologies [20–25].

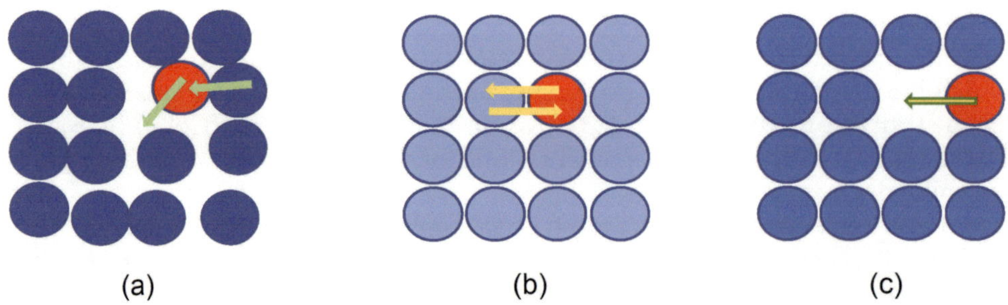

Figure 1.
Atomic movement that results in atomic diffusion. (a) Interstitial diffusion, (b) self-diffusion or inter-diffusion, (c) vacancy diffusion.

2. Diffusion mechanism and coefficient

For the vacancy diffusion mechanism, the probability for any atom in a solid to move is the product of the probability P of finding a vacancy in an adjacent lattice site [25, 26]:

$$P = z\, exp\left(-\frac{G_f}{K_B T}\right) \tag{1}$$

where z is the coordination number (number of atoms adjacent to the vacancy), G_f is the free energy necessary to form the defects, T is the absolute temperature (K), K_B is the Boltzmann constant, and the frequency of jumps (probability of thermal fluctuation needed to overcome the energy barrier for vacancy motion)

$$R_j = V_0\, exp\left(\frac{-\Delta\, G_m}{K_B T}\right) \tag{2}$$

where R_j is the probability of such fluctuation or frequency of jumps, V_0 is an attempt frequency related to the frequency of atomic vibrations, and G_m is the activation free energy for vacancy motion.

Therefore, the diffusion coefficient [27] is

$$D = zV_0 a^2\, exp\left(\frac{-\Delta\, G_m}{K_B T}\right) exp\left(-\frac{G_f}{K_B T}\right) \tag{3}$$

where a is the mean distance between atoms in a crystal lattice.

Eq. 3 can be rewritten as

$$D = D_0\, exp\left(\frac{-\Delta\, G_m - G_f}{K_B T}\right) \tag{4}$$

where D_0 is a parameter of material (both matrix and diffusing species).

Thus, the diffusion coefficient is the measure of the mobility of disusing species:

$$J = -D\left(\frac{d_C}{d_X}\right) \tag{5}$$

where $\frac{d_C}{d_X}$ is the concentration gradients (negative in the direction of diffusion), as shown in **Figure 2**.

Hence, from Eq. 5

$$D = D_0\, exp\left(\frac{-Q_d}{RT}\right) \tag{6}$$

where D_0 is the temperature-independent preexponential (m^2/s), Q_d is the activation energy for diffusion (J/mol or eV/atom), and R is the gas constant (8.31 J/mol K or 8.62 × 10–5 eV/atom K).

By taking the logarithm for Eq. 6, we can get

$$lnD = lnD_0\, \frac{-Q_d}{RT} \tag{7}$$

$$logD = logD_0\, \frac{-Q_d}{2.3RT} \tag{8}$$

From Eq. 8, Q_d the activation energy for diffusion and D_0 independent preexponential can be measured by estimating the $logD_0$ versus 1/T or lnD_0 versus 1/T as the Arrhenius plots (**Figure 3**). **Figures 1** and 2 and **Tables 1** and 2 were taken from Porter and Easterling textbook and *Smithells Metals Reference Book* [1].

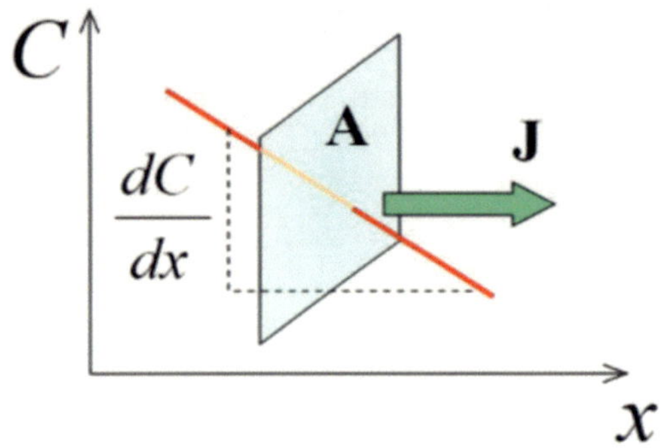

Figure 2.
The slope of particular point on the concentration gradient.

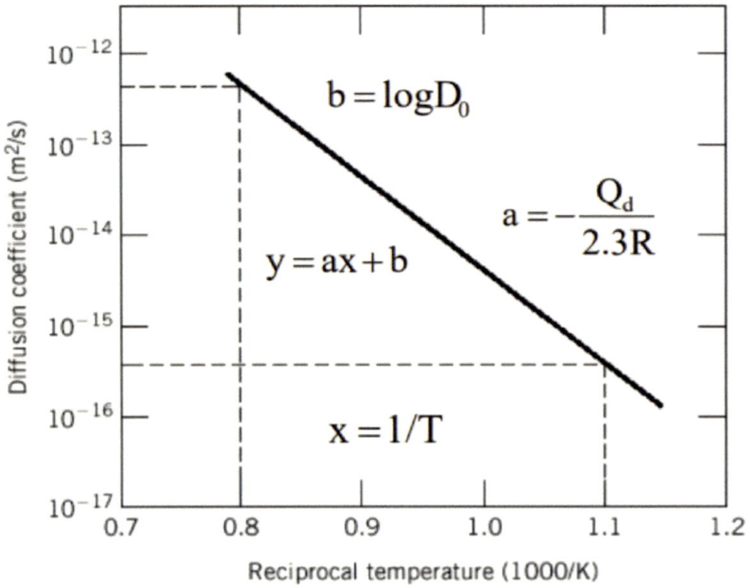

Figure 3.
Arrhenius plots. Q_d as the function of D_o diffusion temperature dependence [1].

Impurity	D_0 = (mm^2/S^{-1})	Q_d = (kJ/mol)
C in BCC Fe	1.1	87
C in FCC Fe	23	138
N in BCC Fe	0.74	77
N in FCC Fe	0.34	145
H in BCC Fe	0.12	15
H in FCC Fe	0.63	43

Table 1.
Examples of the temperature-independent preexponential and the activation energy for diffusion of some atoms in the case of interstitial diffusion mechanism.

Impurity	D_0 = (mm^2/S^{-1})	Q_d = (kJ/mol)
Fe in FCC Fe	65	279
Fe in BCC Fe	410	246
Si in Si	180,000	460
Ni in Cu	230	242

Table 2.
Examples for the temperature-independent preexponential and the activation energy for diffusion of some atoms in the case of vacancy diffusion mechanism.

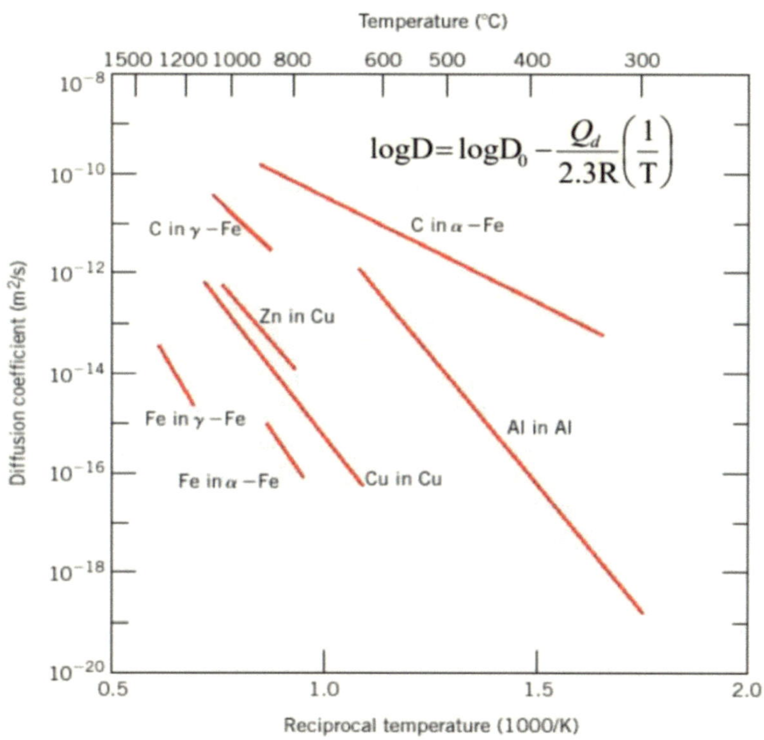

Figure 4.
Logarithm of the diffusion coefficient versus the reciprocal temperature [1].

In Eq. 7, it seems that the vacancy diffusion mechanism is slower than interstitial diffusion, as shown in **Figure 4** and **Tables 1** and 2 (self-diffusion or diffusion of substitutional atoms) [28–31].

From Eq. 6, the big atoms cause more distortion and take more time to diffuse than the smaller atoms during the migration process as we can see from **Tables 1** and 2. Also the diffusion is slower in a close direction and lattices.

3. Fabrication of quantum well intermixing (QWI)

The fabrication of photonic integrated circuits (PICs) by employing an integration of lasers and transparent waveguides on a single epitaxially grown substrate demands the actual understanding and definition of regions possessing different bandgap energy characters. The approach to work out a solution to this problem can be categorized into intermixing and growth approaches. Among the growth approaches, the most popular ones are selective area growth approach and use of a plated substrate to etch-and-regrowth approach [29, 30, 32]. The former one allows for simultaneous epitaxy employing the use of different growth rates, which in turn allows for flexibility toward the growth of quantum wells with varying thicknesses [32–34]. In contrast to the former approach, the latter approach uses different quantum well thicknesses along with subsequent growth of material. Using impurities or vacancies toward the selective partial intermixing of quantum wells provides an alternative approach. The change in the shape of quantum well and thus the transition energies associated occur due to the intermixing of barrier material and quantum well material, which happens during a high-temperature annealing. The capability to identify and define regions that are not to be intermixed and which are to be intermixed acts as the key factor to the viability of the QWI approach. Intermixing method that does not demand epitaxial regrowth is identified to be more cost-effective and potentially simpler [35]. This is the main advantage of the intermixing method. In the following, the means of patterning non-intermixed and intermixed regions along with several QWI approaches are described.

4. Techniques utilized for QW intermixing

There are three techniques of inter-diffusion that are in existence and are widely used. These are the inter-diffusion that is impurity-induced disordering (IID), vacancy diffusion that is IFVD, and laser-assisted disordering (LAD) laser-induced QW intermixing. The first technique uses impurities to accomplish inter-diffusion for the considerable alteration in electrical conductivity and refractive index. Its common utilization is in achieving sideways optical and electrical confinement in semiconductor lasers. In contrast, IFVD does not involve impurities in obtaining inter-diffusion such that there is the conservation of the electrical properties of the diffused QWs. Its typical use is in fabricating tuning LAD technique that has been tested and developed in the last three decades. This method is based on the direct writing of the laser beam into the structure [36–38].

5. Intermixing types and history

Impurity-induced layer disordering (IILD) was the first quantum well intermixing technique to be ever demonstrated. In 1981, the affirmation of disordering of an AlAs-GaAs superlattice employing Zn (Zinc) as the active species was carried out by Laidig et al. [33, 34, 39]. In this affirmation, thermal annealing for several hours was conducted at a temperature of 600°C. As a result, it was identified that in a superlattice, different grades of intermixing can occur according to the anneal conditions used. The fabrication of lasers with emission wavelength (blue shifted) was conducted in 1983 employing this technique [33]. In 1984, it was made possible to laterally define the waveguide of a buried heterostructure employing stripe geometry QW laser devices using IILD [35, 40]. A year after, the first QW laser utilizing transparent facet windows was developed using IILD [33, 41, 42]. The refining of the intermixing method has been happening since then and has currently transformed into one of the best methods which are understood and employed in many commercial products; the most prominent of these includes high-power semiconductor lasers integrated with disordered facet windows. It should be noted that ion implantation can be utilized instead of incorporating impurities (impurities include Si, Mg, or Zn) into the lattice utilizing the process of diffusion. Ion implantation possesses the primary benefit of not automatically incorporating heavy p-type or n-type doping while introducing the reactive species and of having a larger variety of species made available. On the flip side, high implant energies utilized have been identified to cause crystal damage which is not easily removable as in the case with other material systems (e.g., Si material systems). Both of the intermixing processes discussed above rely on the use of impurity atoms to intensify the Al-Ga self-diffusion process by employing different mechanisms. Although discussions and debates still exist around the exact nature relating to the process of intermixing, several experiments and authors have confirmed the unquestionable role of column-III vacancies and column-III interstitial types. A decade ago, methods such as VED, which are impurity-free intermixing methods, gained their popularity since they offered the possibility of intermixing without employing the doping process which prevent the absorption of the free carrier and without crystal damage created by implantation which would, on the other hand, be responsible for scattering loss. An As-rich ambient in a quartz ampoule was employed in the first experiments to prevent crystal surface damage by arsenic out-diffusion [36, 37]. In 1988, the use of an evaporated SiO_2 encapsulant in order to improve the intermixing process was first demonstrated [38]. Soon after, the process of generating vacancies and thus supporting the process of intermixing became possible by employing other dielectrics such as SiON or SiN. In 1993, fluorides (such as SrF or AlF) were identified to prevent QWI in a more effective manner [39]. Essential for the development of optoelectronic devices and instruments, these materials identified allowed for the definition of a certain pattern with dissimilar bandgap energies.

The IILD process makes use of an Ar-based laser beam that is very highly focused in nature. To develop the AlGaAs-GaAs DFQW, the beam of laser marking a wavelength measurement of 488 nanometers (nm) is used to scan the sample which is heterostructure in nature and is also enclosed using a layer of Si-Si_3N_4, which is approximately 90 nm in thickness. The speed of scan employing the laser beam could be marked up to the highest value of 85 pds. The area in which the laser beam interacts will develop an enhanced cylindrical segment identifiable to the range of microns. The process of annealing is then initiated in order to guide the silicon into the required crystal, which will result in the local intermixing of the layers of crystal. On the other hand, to selectively intermix GaInAs over GaInAsP quantum well structure, pulsed photo-absorption-induced disordering (PAID) technique is employed, which was deliberated employing the utilization of time-resolved photoluminescence of high spatial resolution. As a consequence of the above-said process of intermixing, a reduction of approximately two orders of measure in the time of non-radioactive recombination was achieved, which was confirmed from the measurements conducted.

Impurity-induced, impurity-free (dielectric cap), implantation-induced, and laser-induced techniques are some of the QWI techniques that have been advanced. Out of these techniques, the use of impurity-free techniques is strongly advised since optical absorption occurs as a result of the process that the semiconductor waveguide being instituted to dopants which are electrically active in nature. In order to develop vacancies on the group III lattice site, the impurity-free vacancy disordering (IFVD) technique employs the utilization of dielectric caps, which are placed on the semiconductor's exterior surface [3, 7]. The vacancies happen to diffuse through the surface of the semiconductor resulting in solitary atoms bouncing among different lattice sites. Resultantly, it is found that the quantum well intermixes with the adjoining barrier material [34].

6. QW intermixing in industries

From the development of individually addressable laser arrays of higher density to laser-based products within extreme power ranges, monolithic integration platform which is highly innovative and known as quantum well intermixing (QWI) is reshaping methods in which laser diodes are used to solve the ever-increasing optoelectronic requirements. This is particularly important since laser systems which are QWI-enabled are found to deliver far better performance characteristics in factors of power output, luminosity, yield, and dependability.

The QWI is utilized to develop passive waveguides to the interior of the laser cavities adjoining to each facet. It is identified that excellent electro-optical performance is achieved owing to the incorporation of the passive waveguides, especially referring to high-power, single-mode function. An idiosyncratic attribute of this approach is that it allows for the mass production of huge numbers of lasers in parallel, on the very same chip, with very superior efficiency since the passive waveguides are adequately long enough to relax mechanical-related cleaving tolerances. The QWI technologies can be largely employed in many other applications, owing to their farthest versatility. Some of the areas in which the extremely versatile nature of the QWI technology could be utilized to its maximum potential include monolithic photonic integrated circuits (PICs) and in the comprehension of the broad area and stack lasers which provide atypical high-power characteristics and dependability. PICs mainly find their application in broadband optical systems, optoelectronic signal processing systems, microwave photonics, and biophotonics.

QWI gains its importance since it is an integration technique that permits the tampering of the properties of a semiconductor quantum well structure, after its growth. The quantum well intermixing technique combines active and passive components on the very same chip. To manufacture complex laser diodes, laser diode array systems, and photonic integrated circuits (PICs) in a manufacturing environment, intense proprietary QWI technology is utilized. The result of this process is the development of next-generation laser technology which can easily be utilized for a variety of applications [31, 41, 42].

The evolution of the next-generation systems is driven today by the latest innovations in laser diode technology. Intense is providing laser products with far better brightness, improved lifetimes, and increased dependability by employing modernized semiconductor design and patented QWI technology. The ways in which lasers are providing viable solutions to mission-critical problems are revolutionized by the quantum well intermixing method developed by the company, innovatively by producing integrated chips at efficient levels and yields which was unidentified in the industry before.

7. Real applications and fabrication using QWI

In this section, we will summarize some of the laser and light-emitting diode (LED) QWI applications that have been fabricated and tested by our group at the University of Central Florida (UCF) cleanroom facility [14, 43]. These experiments will show the important role of the intermixing and how it can be used for the integrated devices. We will start with the laser followed by the LED.

7.1 Laser diodes

When quickly heated at higher degrees and topped using SiN_x and SiO_yN_x films of various constitutions, quantum well frameworks InGaAsP are interlinked to different degrees. Laser diodes are fabricated with shifted samples of both blue and red, and their output is recorded.

Selective area mixing of semiconductor-based multiple quantum wells (MQWs) could be considered a crucial strategy toward the development of consolidated optoelectronic circuits and instruments. The bandgap energy of the substance can be controlled with stability over a wide spectral range by monitoring the intensity of the intermixing process. The lasers generated on a single monolithic substratum may, therefore, have wavelengths of output which differ widely. The correct combination of the encrusted films may vary the wavelength to either blue or red. The narrow-field semiconductor regrowth procedures have not been very successful in repeatedly producing high-yielding optoelectronic products. Others have documented many techniques for the after-growth combination of QW. In the analysis, we selected a method of induced disorder by impurity-free vacancy that works by rapid thermal annealing (RTA) of QW specimen coated by SiN_x or SiO_yN_x. The range of intermingling could be precisely controlled by changing the dielectric layer capping constitution. Employing this method, we were able to manufacture multiple lasers using a single sample of the InGaAsP multiple quantum well framework, which has been covered by various SiO_yN_x configurations in different parts and annealed at 800°C for 30 s. Slope efficiencies, threshold currents, and laser diodes that are manufactured in the separate section are then carefully defined based on their lasing wavelengths. Such output properties are then juxtaposed with that of the laser diode made employing the primary as-grown multiple quantum well specimen as shown in **Figure 5** [14, 17, 32, 40, 41].

7.1.1 Result

Increasing the ratio between NH_3 and SiH_4 to N_2O during the SiO_yN_x film growth has been found to result in a higher refractive index. It is noted that wavelengths (lasing) of the instruments manufactured on intermixed specimens are identified to be shifted to lower frequencies (red shift). At the same time, the capping film refractive index throughout RTA is higher than the value of 1.95 (refractive index). In comparison, the instruments covered with films having a refractive index lower than 1.95 in value show lasting wavelengths changed blue to higher frequencies. Accordingly, the absolute value of the laser spectrum is experiencing a red shift with a larger ratio in SiN_x film and blue shift with a smaller ratio in SiO_yN_x film as shown in **Figure 6**.

Laser diode made from an as-grown multiple quantum well specimen acted as a base standard and is identified to have a lasing wavelength of 1556 nm. In **Figure 7**, all the fabricated laser diodes are shown with the accompanying spectra. The

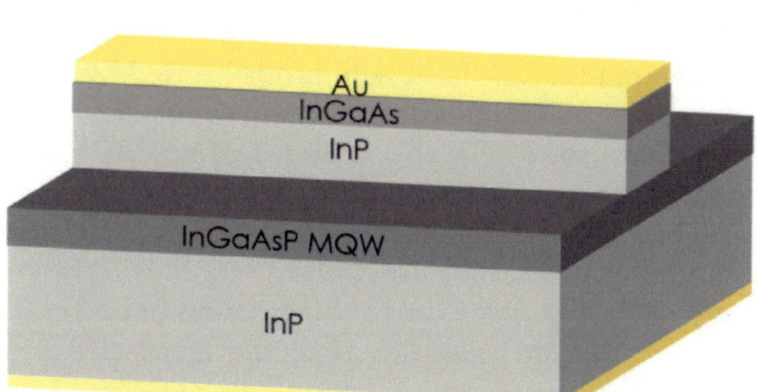

Figure 5.
Schematic of the InGaAsP MQW laser diode with InP substrate as substrate layer and InGaAs as capping layer.

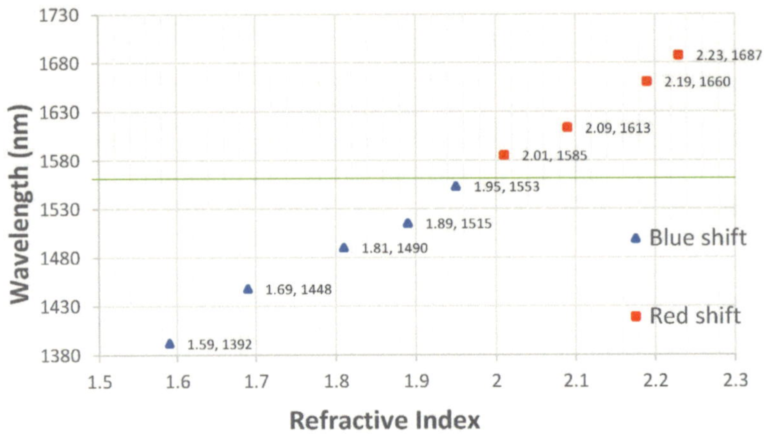

Figure 6.
The absolute values of all lasers' spectrum as a value of the refractive index of the film for different capping layer combinations. The blue shift is associated with SiO_yN_x films, while the red shift is associated with SiN_x.

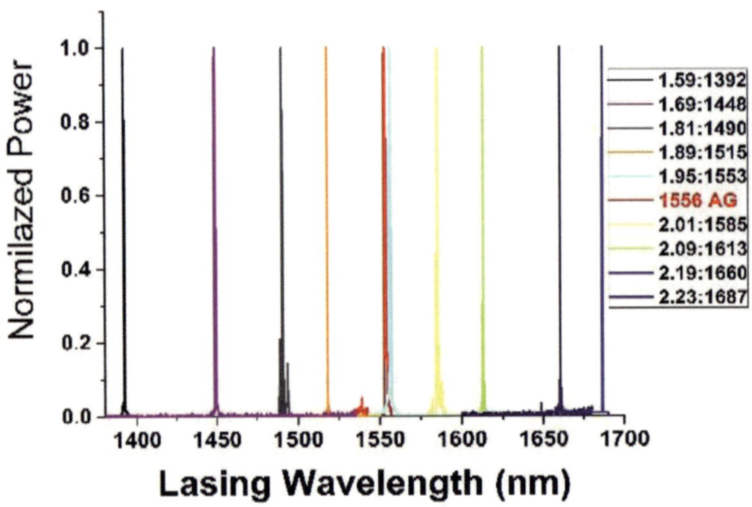

Figure 7.
Laser spectrum of all fabricated devices. It shows the blue and red shifted from the as-grown ones. This figure is taken from [14].

highest blue-shifted wavelength of laser noted is 1392 nm (164 nm change compared to as-grown laser), and 1687 nm (131 nm change as compared to as-grown laser) is the most excellent noted red-shifted wavelength of the laser. Throughout this review, it is discovered that the laser light is not emitted by a system manufactured utilizing a noncapped RTA manufactured multiple quantum well sample. Thus, uncapped regions of the MQW specimen were found to have sustained irreparable harm during thermal annealing [14].

Figure 8 shows the output power curve (L-I curve) for all intermixed laser devices. The laser that fabricated using the most intermixed MQWs had the lowest output power, while the as-grown laser diode has the highest output power. Therefore, as we intermixed more, we create more losses that affect the device efficiency.

7.2 LED and modulators on InGaAsP

Using a controllable technique for the red and blue shifting of bandgap energy of the quantum well, we were able to develop LED sources that reach a broad

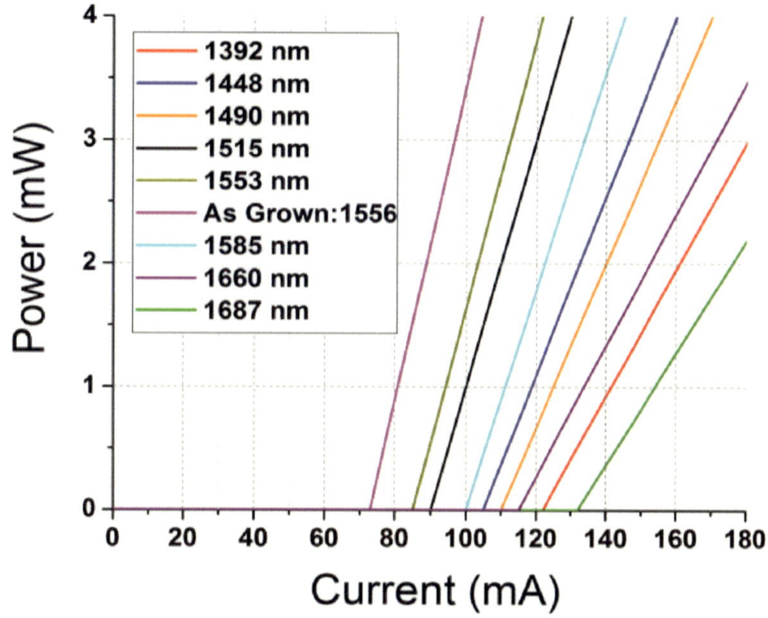

Figure 8.
The fabricated lasers diode as function of threshold current L-I curve.

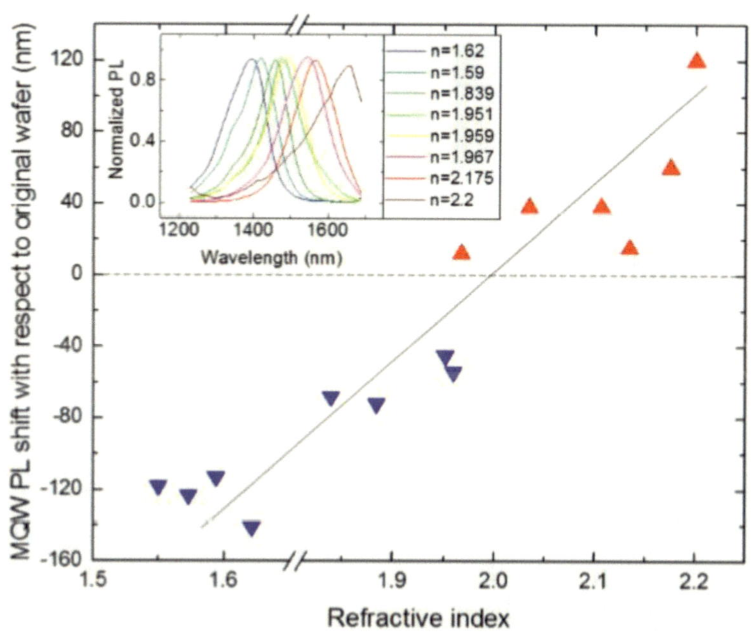

Figure 9.
Measured absolute value of the PL shift of the RTA-treated samples from that of the as-grown wafer for different dielectric film capping, in respect to the refractive index of the film. The blue shift is associated with SiO_yN_x films for different ratios of NH_3/N_2O, while the red shift refers Si-rich compositions. The inset shows the absolute PL spectrum for selected data points. This figure and caption were taken from [43].

frequency spectrum along with all-optical modulator intensity instruments. Through using an impurity-free vacancy diffusion method, they show bandgap adjustment of multiple quantum well structures of InGaAsP. By utilizing SiO_2, SiO_yN_x, and SiN_x capping layers, and by regulating the related oxygen and nitrogen content, a significant modification of the bandgap energy toward the red and blue portions of the spectrum is identified. The subsequent degree of tuning, with band-to-band wavelength emissions of up to 120 nm red shift and 140 nm blue shift, was analyzed using photoluminescence at room temperature, following the emission spectra acquired from LED semiconductor instruments manufactured on this framework. The intensity modulator instruments are made along with compatible LED sources for the chosen frequency, designed to achieve minimal material losses and modulation of residual amplitude as shown in **Figure 9** [43].

The fabricated LED has been integrated with transparent intensity modulator as shown in **Figure 10**. The intensity modulator is based on a Mach-Zehnder

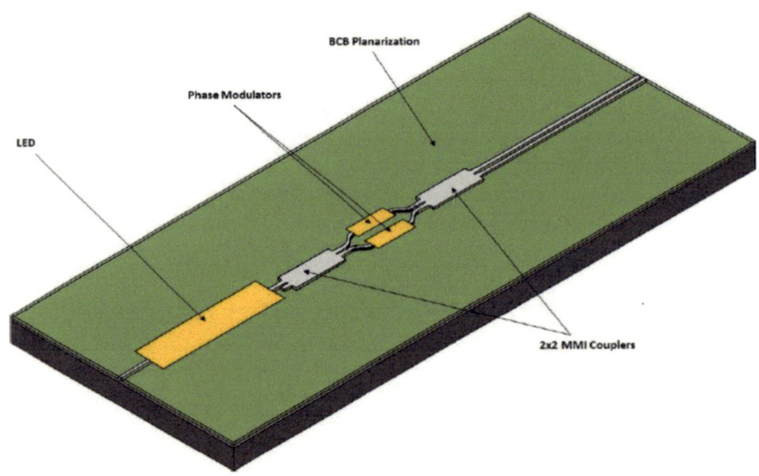

Figure 10.
Schematic for the integrated LED with MZI intensity modulator [43].

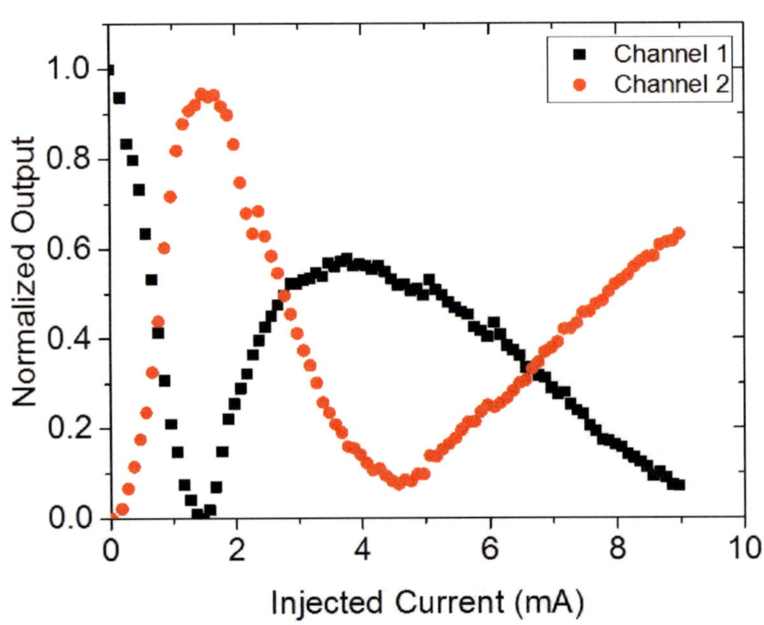

Figure 11.
The output power as function of injected current for the integrated device.

interferometer (MZI) where the phase control is achieved by injecting electrons into the core of the waveguide.

As the light source from the LED passes through the MM-MZI device, the outpower changes. The result has been recorded and evaluated as the function of the passing current as shown in **Figure 11**.

8. Conclusion

In this chapter, we have studied and compared the different methods for diffusion of atoms into both surface and internal layers. Also, we have shown the variety of QWIs that change and modify the refractive index and energy bandgap of QW's structures. There are several QWI techniques accessible, and each technique has specific characteristics that are useful under various circumstances. Very likely, more than one process will be used to produce a semiconductor chip. Among the techniques used for this purpose, owing to their capacity to preserve the electrical properties of the QW structure and its strong selectivity throughout the spatial domain, triggered disordering of MQWs by using impurity-free vacancy diffusion process gained much interest. A selective area QWI procedure is used that includes vacancy diffusion via the fast-thermal strengthening of the sample which is capped by silicon dioxide or different silicon oxynitride coatings. Prior to the fast-thermal annealing of the specimen, it is identified that the bandgap energy of the intermixed QW system can be efficiently managed by varying the dielectric capping film composition. As an illustration, for laser, we displayed the implications of intermixing of laser diodes based on InGaAsP QWs. By adjusting the proportion of mixed films, it was possible to adjust the lasing wavelength to the red or blue shift regions. Using an impurity-free vacancy diffusion method, we illustrated bandgap adjustment of several quantum well structures of InGaAsP, which was then used for the LED applications. By utilizing SiO_2, SiO_yN_x, and SiN_x capping films and by regulating the corresponding oxygen and nitrogen levels, a significant alteration of the bandgap energy toward the red and blue segments of the spectrum was achieved. The resultant level of adjustment was noted, red shift up to 120 nm and band-to-band blue shift of 140 nm.

Acknowledgements

I want to give special thanks to my professor and advisor Prof. Patrick Lickamwa for his help and advice through all my PhD. Also, I would like to thank UCF and CREOL for letting me use their cleanroom facility. Finally, many thanks go to King Abdulaziz City for Science and Technology for their supports.

Author details

Thamer Tabbakh
King Abdulaziz City for Science and Technology, National Center for Nanotechnology and Semiconductor, KACST, Riyadh, Saudi Arabia

*Address all correspondence to: t.tabbakh@gmail.com

IntechOpen

© 2020 The Author(s). Licensee IntechOpen. This chapter is distributed under the terms of the Creative Commons Attribution License (http://creativecommons.org/licenses/by/3.0), which permits unrestricted use, distribution, and reproduction in any medium, provided the original work is properly cited.

References

[1] Gale WF, Totemeier TC. Smithells Metals Reference Book. Elsevier; 2003

[2] Hilliard J, Averbach B, Cohen M. Self and interdiffusion in aluminum-zinc alloys. Acta Metallurgica. 1959;7(2):86-92

[3] Bachrach R, Bauer R. Surface reactions and interdiffusion. Journal of Vacuum Science and Technology. 1979;**16**(5):1149-1153

[4] Pape IJ, Wa PLK, David JPR, Claxton PA, Robson PN, Sykes D. Diffusion-induced disordering of Ga/sub 0.47/In/sub 0.53/As/InP multiple quantum wells with zinc. Electronics Letters. 1988;**24**(15):910-911

[5] Aimez V et al. Low-energy ion-implantation-induced quantum-well intermixing. IEEE Journal of Selected Topics in Quantum Electronics. 2002;**8**(4):870-879

[6] Stephenson GB. Deformation during interdiffusion. Acta Metallurgica. 1988;**36**(10):2663-2683

[7] Koch TL et al. Tapered waveguide InGaAs/InGaAsP multiple-quantum-well lasers. IEEE Photonics Technology Letters. 1990;**2**(2):88-90

[8] O'neill M et al. Reduction of the propagation losses in impurity disordered quantum well waveguides. Electronics Letters. 1990;**26**(19):1613-1615

[9] Marsh JH. Quantum well intermixing. Semiconductor Science and Technology. 1993;**8**(6):1136

[10] Charbonneau S et al. Quantum-well intermixing for optoelectronic integration using high energy ion implantation. Journal of Applied Physics. 1995;**78**(6):3697-3705

[11] Hamoudi A et al. Cation interdiffusion inter-in InGaAsP/InGaAsP multiple quantum wells with constant P/As ratio. Applied Physics Letters. 1995;**66**(6):718-720

[12] Oh Y, Kang T, Kim T.Photo-luminescence and photoreflectance from GaAs/AlAs multiple quantum wells. Journal of Applied Physics. 1995;**78**(5):3376-3379

[13] Poole P et al. Defect diffusion in ion implanted AlGaAs and InP: Consequences for quantum well intermixing. Journal of Applied Physics. 1995;**78**(4):2367-2371

[14] Tabbakh T, LiKam WP. Blue and red shifted, partially intermixed InGaAsP quantum well semiconductor laser diodes. In: 2017 IEEE Photonics Conference (IPC)

[15] Alferness R et al. Broadly tunable InGaAsP/InP laser based on a vertical coupler filter with 57-nm tuning range. Applied Physics Letters. 1992;**60**(26):3209-3211

[16] Sun H et al. Characterization of selective quantum well intermixing in 1.3 μm GaInNAs/GaAs structures. Journal of Applied Physics. 2003;**94**(3):1550-1556

[17] Tabbakh T, LiKam WP. Dual wavelength single waveguide laser diode fabricated using selective area quantum well intermixing. Optik. 2017;**140**:592-596

[18] Teng J et al. Dual-wavelength laser source monolithically integrated with Y-junction coupler and isolator using quantum-well intermixing. IEEE Photonics Technology Letters. 2000;**12**(10):1310-1312

[19] Leon R et al. Effects of interdiffusion on the luminescence of InGaAs/GaAs quantum dots. Applied Physics Letters. 1996;**69**(13):1888-1890

[20] Alahmadi Y, LiKam WP. Effects of selective area intermixing on InAlGaAs multiple quantum well laser diode. Semiconductor Science and Technology. 2019;**34**(2):025010

[21] Beall R et al. Gallium arsenide and related compounds. In: 1988 Inst. Phys. Conf. Ser. 96. Bristol: Institute of Physics; 1989

[22] Forouhar S et al. InGaAs/InGaAsP/InP strained-layer quantum well lasers at approximately 2 μm. Electronics Letters. 1992;**28**(15):1431-1432

[23] May-Arrioja D et al. Intermixing of InP-based multiple quantum wells for integrated optoelectronic devices. Microelectronics Journal. 2009;**40**(3):574-576

[24] May-Arrioja D et al. Intermixing of InP-based multiple quantum wells for photonic integrated circuits. In: AIP Conference Proceedings. American Institute of Physics; 2008

[25] Zucker J et al. Large blueshifting of InGaAs/InP quantum-well band gaps by ion implantation. Applied Physics Letters. 1992;**60**(24):3036-3038

[26] Lazarus D. Diffusion in metals. In: Solid State Physics. Elsevier; 1960. pp. 71-126

[27] Shewmon P. Diffusion in Solids. Springer; 2016

[28] Jost W. Diffusion in Solids, Liquid, Gases. New York: Academic Press Inc; 1960. p. 73

[29] Chang L, Koma A. Interdiffusion between GaAs and AlAs. Applied Physics Letters. 1976;**29**(3):138-141

[30] Lee M et al. Intermixing behavior in InGaAs/InGaAsP multiple quantum wells with dielectric and InGaAs capping layers. Applied Physics A. 2001;**73**(3):357-360

[31] Liu C-C et al. Intermixing in InGaAs/AlGaAs quantum well structures induced by the interdiffusion of Si impurities. 2020

[32] Tabbakh T, LiKam WP. Intermixed InGaAsP MQW tunable laser diode suitable for probing surface plasmon resonance optical sensor. In: Nanoengineering: Fabrication, Properties, Optics, and Devices XV. International Society for Optics and Photonics; 2018

[33] Laidig W et al. Disorder of an AlAs-GaAs superlattice by impurity diffusion. Applied Physics Letters. 1981;**38**(10):776-778

[34] Holonyak N Jr, Laidig WD, Camras MD, Coleman JJ, Dapkus PD. Applied Physics Letters. 1981;**39**:102

[35] Sasaki T, Kitamura M, Mito I. Selective metalorganic vapor phase epitaxial growth of InGaAsP/InP layers with bandgap energy control in InGaAs/InGaAsP multiple-quantum well structures. Journal of Crystal Growth. 1993;**132**(3-4):435-443

[36] Qiao Z et al. Monolithic fabrication of InGaAs/GaAs/AlGaAs multiple wavelength quantum well laser diodes via impurity-free vacancy disordering quantum well intermixing. IEEE Journal of the Electron Devices Society. 2017;**5**(2):122-127

[37] Skogen EJ et al. Monolithically integrated active components: A quantum-well intermixing approach. IEEE Journal of Selected Topics in Quantum Electronics. 2005;**11**(2):343-355

[38] Ralston J et al. Room-temperature exciton transitions in partially intermixed GaAs/AlGaAs superlattices. Applied Physics Letters. 1988;**52**(18):1511-1513

[39] Beauvais J et al. Suppression of bandgap shifts in GaAs/AlGaAs

quantum wells using strontium fluoride caps. Electronics Letters. 1992;**28**(17):1670-1672

[40] Tabbakh T, LiKam WP. Quantum well intermixed tunable wavelength single stripe laser diode in active photonic platforms IX. In: International Society for Optics and Photonics. 2017

[41] Tabbakh T, LiKam WP. Tunable laser diode using partially intermixed InGaAsP multiple quantum well. In: Laser Applications in Microelectronic and Optoelectronic Manufacturing (LAMOM) XXIII, Vol. 10519. International Society for Optics and Photonics. 2018

[42] Kumar R et al. Realization of high-quality InGaAs/GaAs quantum dot growth on Ge substrate and improvement of optical property through ex-situ ion implantation. Journal of Luminescence. 2020:117208

[43] Aleahmad P et al. Controllable red and blue bandgap energy shifted LEDs and modulators on InGaAsP quantum well platform. In: Integrated Optics: Devices, Materials, and Technologies XX. International Society for Optics and Photonics. 2016

Chapter 7

Development and Characterization of High-Quality HfO_2/InGaAs MOS Interface

Sukeun Eom, Min-woo Kong and Kwang-seok Seo

Abstract

The scope of this chapter is to introduce a highly efficient HfO_2 atomic layer deposition (ALD) process with superior interface defect characteristics that can be applied on high-mobility III-V substrates. For a long time, the major academic research of III-V metal-oxide-semiconductor (MOS) studies was mainly oriented on searching for the suitable high-k dielectric, and among the reported high-k/III-V MOS studies, Al_2O_3 and AlN have demonstrated the most promising results. However, usually, the dielectrics with higher dielectric constant suffered from more defective interface quality including the HfO_2, which should be overcome to meet the intensive operation voltage scaling requirements. In order to protect the interface of the HfO_2/III-V MOS, the exposed III-V surface has to be carefully treated before, while, and after the whole high-k deposition process. For this purpose, the effect of isopropyl alcohol precursor and in situ cyclic nitrogen plasma treatment on the HfO_2 ALD process at III-V substrates was thoroughly investigated. Remarkable interface state density levels with strong inversion behavior were achieved, which have not been observed at the previous HfO_2/InGaAs studies. Also, detailed analysis of the interface characteristics was investigated to broaden the understanding of the improvement phenomenon.

Keywords: high-k oxides, hafnium oxide (HfO_2), atomic layer deposition (ALD), III-V channel, indium gallium arsenide (InGaAs), metal-oxide-semiconductor (MOS)

1. Introduction

Over the past decades, the semiconductor foundry business has gone through a dynamic transformation. Recently, the foundries are leading the process development race at 10 nm [1, 2] and even to 7 nm [3, 4] and will continue to do so. However, the traditional physical scaling of advanced MOSFETs in conjunction with Dennard's scaling rules has become extremely challenging as to increase the drive currents for faster switching speeds at lower supply voltages is largely at the expense of large leakage current in extremely scaled device [5]. As a result, even with the huge R&D investments, the semiconductor firms gradually lagged the advertised on-chip feature sizes demonstrated in the scaling roadmap, and finally the end of Moore's law has been declared with the end of the 2016 International Technology Roadmap of Semiconductors (ITRS) [5, 6]. Also, the emergence of internet of things (IoT) and big data applications has driven a necessity of abundant

IntechOpen

computing and memory resources that requires always-on and high-performance ultralow-power devices to generate data instantly. Several device architectures and novel materials based on both analytical and experimental academic research were proposed in the metal-oxide-semiconductor field-effect transistor (MOSFET) technology. Among the viable technologies, the compound semiconductor especially the III-V materials have stood out to be a promising channel candidate for the future highly scaled CMOS application.

The light effective mass of III-V materials compared to the Si even in the highly strained case leads to a higher electron mobility and a higher injection velocity, which should translate into a great turn-on performance even at a lower operation voltage (V_{DD}) level down to 0.5 V. Moreover, there is already a mature industry that uses III-V high electron mobility transistors (HEMTs) for high-frequency applications [7, 8], and it provides excellent techniques such as InGaAs and InAs quantum well (QW) FETs [9, 10]. However, most of these III-V compound semiconductors have smaller bandgaps, which have great impact on the band-to-band tunneling leakage currents. In addition, according to Yan's model [11], the higher permittivity of these materials may worsen the short channel effects (SCE). In spite of the demerits that may limit the scalability, the benefits are much more attractive which makes the III-V channel technology a powerful beyond CMOS solution. However, the use of III-V compound semiconductors has been reluctant to the industry because of its high-cost manufacturing process and CMOS-incompatible process. Naturally, it brought out a strong motivation of research of III-V hetero-integration on a Si platform. The main obstacle of III-V on Si integration research is that as huge lattice constant mismatch exists between those two materials, growing epitaxial films directly on Si without defect is difficult [12]. Accordingly, different approaches have been developed, and among them, direct wafer bonding [13, 14] and aspect ratio trapping (ART) [15, 16] technologies have projected the most promising results.

Consequently, the remaining issue toward the practical realization of III-V materials is its defective interface quality which has been the major drawback compared to Si [17–21]. The poor native oxide quality compared with SiO2 is challenging even more with III-V materials. The III-V compound semiconductors are typically composed of binary, ternary, or even quaternary material by covalent bonding, and more complex elements mean a much richer population of possible oxides for the III-V materials [22]. These native oxides are not thermodynamically stable and very leaky that rise serious issues of creation of significant surface states on the oxide-semiconductor interface and huge trap-assisted gate leakage current [17, 19]. At the early stage of research, GaAs MOSFET suffered from high density of interface states (over 3 orders compared to Si) hindering inversion mode operation.

In order to overcome the defective interface problem, many research groups conducted extensive research effort with a search for a perfect gate dielectric that suits the III-V substrate [23, 24]. The study of atomic layer deposition (ALD) high-k dielectric led to a successful integration of high-k gate dielectrics on III-V substrate, and recent research is mainly focused on the development and interface characterization of ALD high-k and III–V compound semiconductor. To evaluate the objective III-V metal-oxide-semiconductor (MOS) characteristics, it is important to understand the trapping mechanism and know what kind of measurement is required. For Si, the primary defects are the well-known P_b centers, which are dangling bonds at the immediate interface with the dielectric [25]. However, for the III-V material, the anti-sites and interstitials are the critical defect centers [17], and the small DOS of the III-V materials is also a weak point [26]. These differences lead to different trapping mechanism, and unlike Si MOS, the III-V MOS gate stack often exhibits a particular C-V phenomenon typically known as the frequency dispersion effect [27].

The features of the frequency dispersion effect are threefold. First, large inversion-like hump occurs even at high measurement frequency, which could not be an actual inversion characteristic theoretically. Secondly, the C-V curve horizontally shifts to the negative direction as the measurement frequency decreases. Finally, the accumulation capacitance increases as the measurement frequency decreases.

The large interfacial trap densities (D_{it}) that reside within high-k dielectric and III-V substrate are mostly responsible for the explained features [28]. The high D_{it} especially the near mid-gap states act as generation recombination centers that attribute to the inversion hump phenomenon in the weak inversion regime. In addition, the large donor-like D_{it} near the conductance band (for n-type substrate) induces a substantial surface charge that needs to be compensated by larger gate biases resulting in a horizontal shift in the C-V curve. Detailed discussions are well explained through both theoretical and experimental research [27, 28]. The accumulation capacitance increase, however, is quite difficult to be explained only by the interface traps. There have been numerous publications on this particular accumulation dispersion behavior, and discussion led to an explanation of a carrier transport model from the crystalline semiconductor into the border traps, which are defects within the bulk of the dielectric [29]. The capture and emission process occur at border traps with the interaction of conduction band electrons resulting in discrepancy of accumulation capacitance, and the thermal barrier in capture process is responsible for the strong temperature dependency.

Among the reported high-quality insulator/InGaAs interface studies, the direct deposition of hafnium oxide (HfO_2) on InGaAs substrate has generally led to poor electrical characteristics, and there are only few studies aimed at improving the intrinsic HfO_2/InGaAs interface quality [30, 31]. These studies also target only in pretreatments, which is vulnerable during oxide deposition. Meanwhile, O_3 and H_2O are the most common oxidants employed in HfO_2 ALD. However, one of the disadvantages of H_2O-based ALD is high-concentration hydroxyl groups in the films, which degrades the dielectric interface during the post deposition annealing process [32]. In addition, sufficiently long purge time is needed because H_2O tends to physisorb on the surface strongly, especially at low temperature. To solve this problem, O_3 is used as one of the most promising alternative oxidants in ALD process, due to its strong oxidization and high volatility. However, O_3 is known to oxidize the III-V surface during the initial deposition cycles which will neglect the prior surface treatments that easily cause the formation of inferior native oxides [33]. The excess interfacial oxidation of the InGaAs surface initiated by the use of ozone is widely reported in the previous studies. H_2O oxidant also is not totally free from surface oxidation [34]. Therefore, the research on alternative oxidation sources is necessary for the HfO_2/InGaAs MOS studies to make the effort made in the pretreatment studies work.

2. Development of IPA-based ALD HfO_2 on n-type InGaAs substrates

Looking into the oxidant candidates, isopropyl alcohol (IPA) is known to be irresponsive to the semiconductor surface during the initial ALD cycles [35], and as most pretreatment studies are aimed at removing the native oxides of the III-V surface, the IPA oxidant will be able to efficiently suppress the surface oxidation after the pretreatment process.

In order to study the effect of using IPA oxidant, O_3 was used as the reference to compare. The basic cycle of the HfO_2 deposition is consisted of a TEMAH precursor pulse and an oxidant (O_3 or IPA) exposure with N_2 purging process between the precursor injection and oxidant process. The temperature of the IPA precursor was maintained at 4°C. The vapor pressure of IPA at 4°C is around 10 mmHg, which is four times smaller than that at the room temperature [36]. It is important to control the excessive vapor pressure because it leads to a longer purge time, which disables an efficient ALD cycle.

The ALD characteristics of HfO_2 using O_3 and IPA oxidants are shown in **Figure 1**. Oxidant pulse times were 1 and 3 s for O_3 and IPA, respectively, which

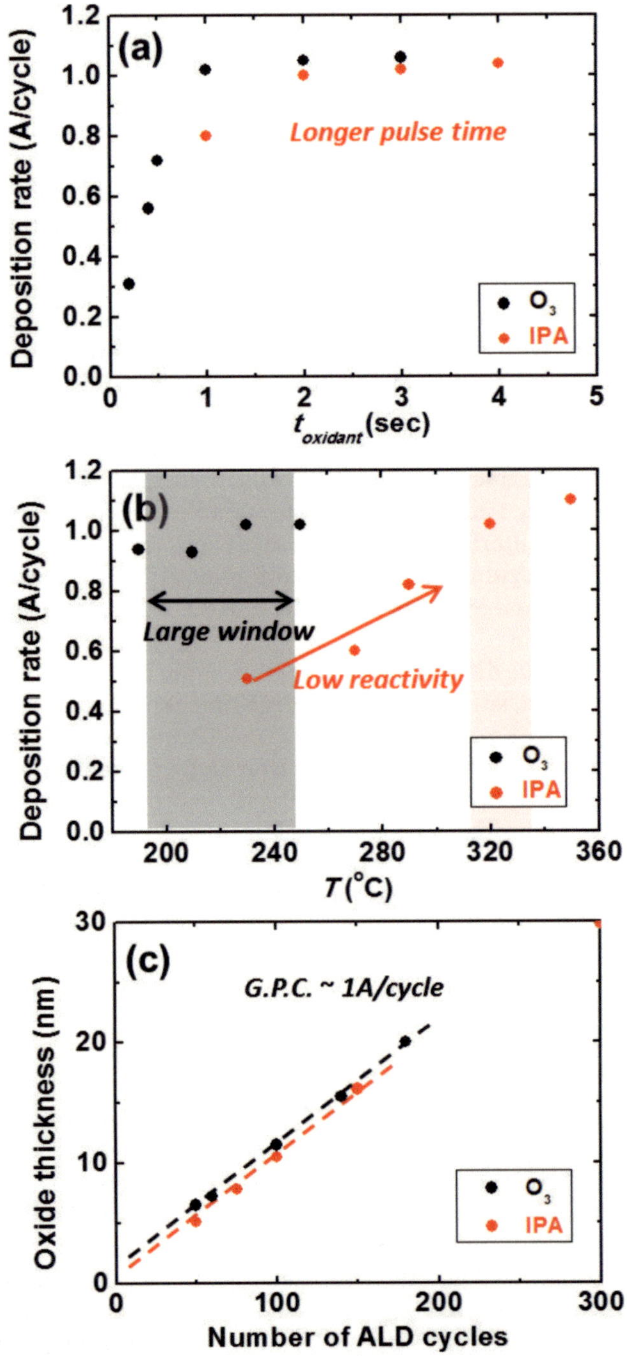

Figure 1.
Comparison of the O3- and IPA-based HfO2 ALD characteristics: (a) deposition rate vs. oxidant time, (b) deposition rate vs. deposition temperature, and (c) growth per cycle rate.

were chosen to meet the saturation requirement of ALD. Both oxidants had similar saturated deposition rate of 0.1 nm/cycle. Noticeable difference was observed in the temperature windows of oxidant type. While stable deposition rate of O_3 oxidant was maintained in a large temperature range, saturated deposition rate of IPA oxidant was only observed in a small temperature range around 320°C. In low temperatures, low deposition rate is due to insufficient reaction which is originated from low reactivity of IPA. Also, in high temperatures above 320°C, thermal decomposition of Hf precursor occurs, and it hinders the self-limiting characteristics of ALD. Therefore, the deposition temperatures of HfO_2 ALD were chosen to be 230 and 320°C for O_3 and IPA, respectively. Moreover, the film thickness per ALD cycles is presented. It is observed that the linear deposition rate per cycle is obtained for both oxidants and a thicker interface layer thickness appears to be existed for the O_3 oxidant due to is strong reactivity (**Figure 2**).

Based on the ALD characteristics, the HfO_2/Si MOS capacitors are fabricated on the Si substrate. All samples underwent standard Si cleaning steps that consisted of SPM- and HF-based cleaning and 400°C 10 min annealing after the dielectric deposition. The C-V and forward gate leakage characteristics are measured and discussed. First of all, the C-V hysteresis difference is notable. As anticipated, the C-V hysteresis significantly decreases by employing the IPA oxidant. Powerful

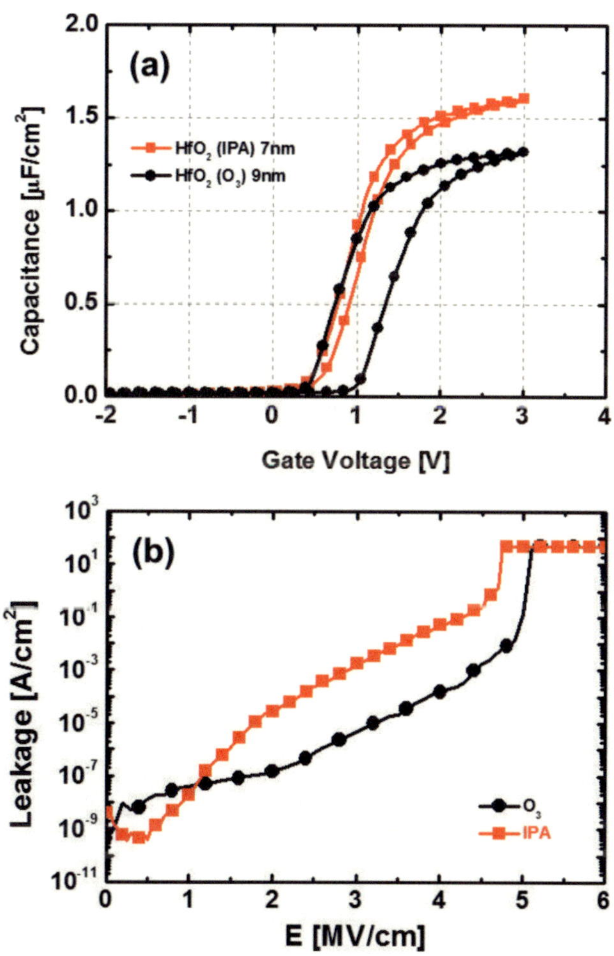

Figure 2.
Comparison of O_3- and IPA-based HfO_2/Si MOS capacitors: (a) C-V and (b) leakage-E plot.

oxidation ability of ozone may induce undesired interfacial oxide at the Si interface forming defective hafnium silicate leading to a large hysteresis, while IPA-based HfO_2 appears to be negligible on this effect [37]. The dielectric constants of IPA-based and O_3-based HfO_2 extracted by the thickness series method are 19.4 and 17.6, respectively [38]. While the C-V results report promising potential of IPA oxidant in ALD HfO_2, the leakage properties suggest a different aspect. Leaky forward gate leakage especially in the medium gate voltage range of the IPA-based HfO_2 is presented compared to the O_3-based HfO_2. It is well known that at this gate bias range, the dominant leakage mechanism is by the Poole-Frenkel tunneling, which is a conduction method of electron tunneling from a metal electrode to traps in a nearby insulator layer, followed by detrapping of the electrons from the traps by virtue of a lowered potential well due to an applied electric field [39]. It usually implies the bulk quality of dielectric; in short, the larger the leakage in this E-field is, the more inferior the gate insulator is. It is speculated that by using the IPA oxidant, the bulk quality may be inferior than using the O_3 oxidant in ALD HfO_2. This might affect the further scaling down potential and the border trap density in ALD HfO_2 application on InGaAs substrate [40].

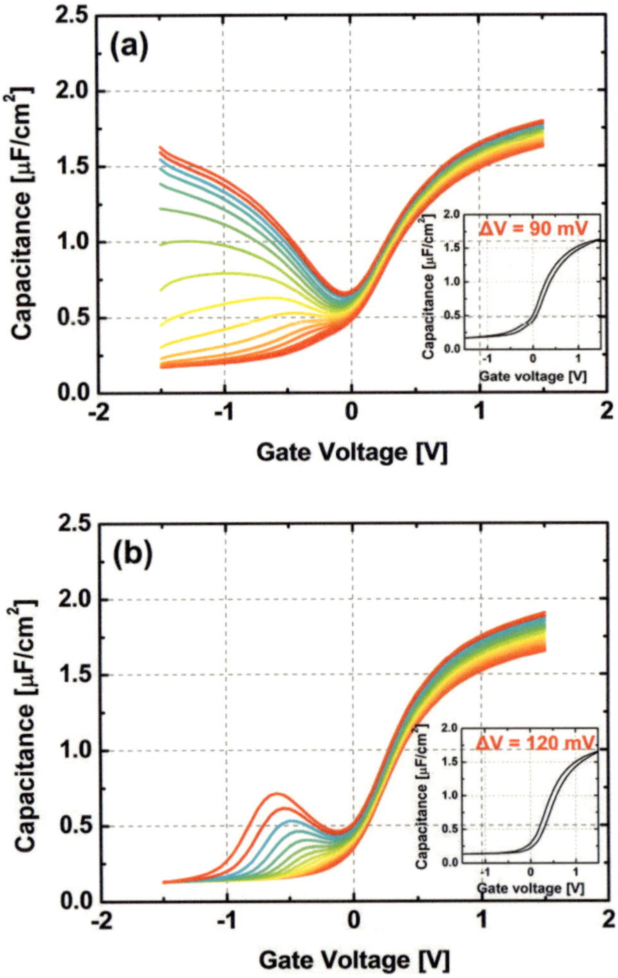

Figure 3.
Multifrequency C-V responses of (a) O_3- and (b) IPA-based HfO_2/In0.53Ga0.47As MOS capacitors; insets are the hysteresis at 1 MHz.

DOI: http://dx.doi.org/10.5772/intechopen.92424

By using the developed O_3- and IPA-based HfO_2 dielectrics, HfO_2/InGaAs MOS capacitors were successfully demonstrated [41]. The multifrequency (1 kHz–1 MHz) C-V characteristics of HfO_2/n-$In_{0.53}Ga_{0.47}As$ MOS capacitors using the O_3 and IPA oxidants are presented in **Figure 3**. The C-V curves of the O_3-based HfO_2 ALD showed a large inversion hump in the negative bias range, which is attributed to the large density of interface defect states near the mid-gap trap level. In contrast, those of the IPA-based HfO_2 ALD showed a notable suppression of the inversion hump behavior. In addition, by employing the IPA oxidant, the effective oxide thickness (EOT) has decreased. We hypothesize that the reduced inversion hump and decrease of the EOT originate from the suppression of unintentional interfacial oxides by the use of the IPA oxidant. Detailed material characteristics analysis was conducted and proved the hypothesis to be convincing [41]. To our knowledge, it is the first successful demonstration of HfO_2 deposition using IPA at InGaAs substrate.

Despite the advantages of using the IPA oxidant, frequency dispersion at the accumulation region slightly increased from 3.3 to 4.7% per decade. In **Figure 4**, these values were used to estimate the border trap densities (N_{bt}) by using a distributed bulk-oxide trap model, and increased N_{bt} of 1.1×10^{20} cm^{-3} eV^{-1} was extracted compared to 6.7×10^{19} cm^{-3} eV^{-1}. Also, larger C-V hysteresis and severely degraded

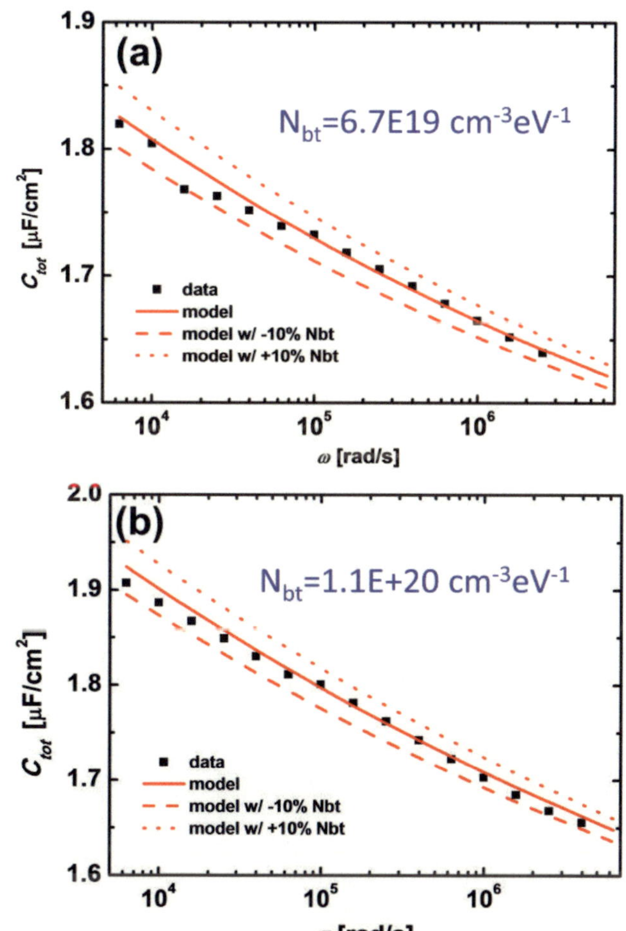

Figure 4.
Border trap estimation of HfO_2/InGaAs MOS capacitors by using the distributed oxide bulk trap model (a) O_3-HfO_2 and (b) IPA-HfO_2.

leakage currents at positive bias are noticed. Based on these results, an inferior quality of the HfO_2 film for using IPA oxidant was predicted which should be resolved for reliable use of the IPA oxidant.

In order to improve the weak IPA-based HfO_2 bulk quality, the study of origin in HfO_2 defect is necessary. One of the main concerns in the replacement of SiO_2 to HfO_2 is that compared to SiO_2, HfO_2 generally suffers from high defect densities leading to several issues such as large carrier trapping, mobility degradation due to coulombic scattering in the channel surface, and threshold voltage shifts in gate stress conditions [42]. To be specific, the threshold voltage shift issue was not a new phenomenon that suddenly happened with use of HfO_2. In immature SiO_2 MOSFETs, it is widely known that the extrinsic contaminations in SiO_2 with alkali ions induce this similar phenomenon [43]. However, with HfO_2, it appeared to be caused by the high defect concentrations, which originated from a more fundamental problem, not an extrinsic defect. Consequently, many researches were devoted to HfO_2 physical model simulation in order to identify the type of defects and their energy levels, and by these physical studies, researchers hoped to learn how the deposition and processing conditions can be optimized to minimize these defect origins [42, 44, 45].

Based on computational calculations, it is identified that oxygen vacancies in HfO_2 are both the principal trap and main cause of the discussed issues, and its formation energy and energy levels were also calculated [44]. Hence, in order to reduce the defect densities, experiments regarding deposition and post processing conditions were aimed to remove or passivate these defects, with an oxygen-rich ambient. However, in many cases, it only worked to some extent and led to new issues of excessive oxidation leaving oxygen interstitials and oxygen diffusion to the interface [46].

Additionally, due to the low density of states of III-V semiconductors, III-V substrates are heavily influenced to border traps that could severely worsen the device performance resulting in poor reliability properties. Therefore, not only the interface but also the bulk characteristics of HfO_2 should be considered in III-V MOS studies, and improvement in both qualities is definitely important.

One of the most effective methods to improve the inherent properties of HfO_2 is the incorporation of nitrogen to passivate oxygen vacancies, and it has been extensively utilized in many recent studies [47–51]. Significant improvement in the electrical characteristics of various high-k gate dielectrics by nitrogen incorporation has been demonstrated, and it is found that interfacial layer growth is effectively suppressed [49] and there is lower boron penetration with nitrogen incorporation [51]. Also, lower leakage current density in HfO_xN_y is widely reported due to suppression of oxygen vacancy traps [50]. It has been reported that nitrogen incorporation in HfO_2 can be achieved by several methods mostly by nitrogen ambient plasma-based nitridation [47, 50] or ammonia (NH_3) ambient high-temperature annealing treatment [52, 53]. For Si-based MOS studies, the later approach is known to be very powerful for achieving good uniformity of nitrogen incorporation and excellent interface quality due to the absence of plasma damage. However, in order to successfully apply nitrogen incorporation technology on III-V substrate, the low thermal budget of III-V compound semiconductor always has to be considered, and high-temperature annealing treatment should be ruled out for nitrogen incorporation study in III-V MOS. In the other hand, although plasma-based nitridation technology offers low thermal budget capacity, most studies generally suffers from several issues such as nonuniform nitrogen distribution throughout dielectric, plasma-induced damage due to high-power plasma for dielectric penetration, and high energy potential nitrogen species substituting well-combined Hf-O bonds. Post deposition plasma treatments have recently been suggested for InGaAs MOS

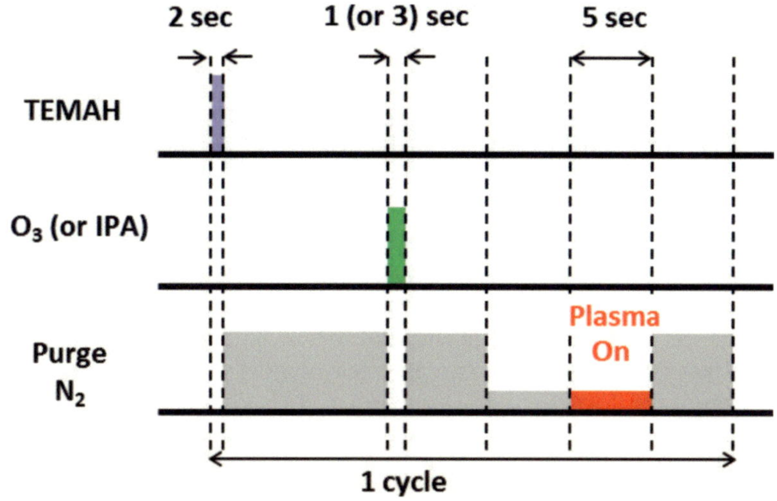

Figure 5.
ALD cycle sequence of the developed HfO$_x$N$_y$ processes on InGaAs substrates.

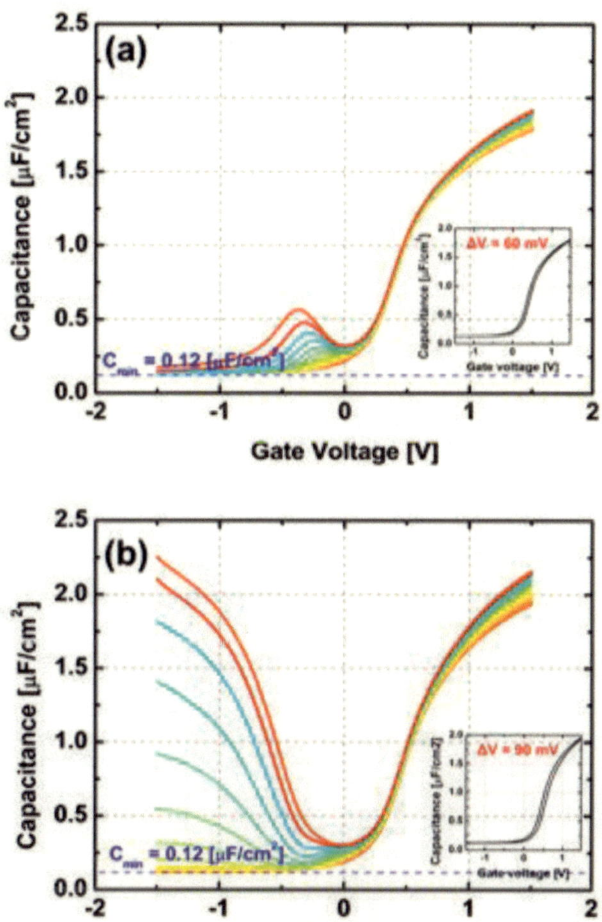

Figure 6.
Multifrequency C-V responses of (a) O$_3$- and (b) IPA-based HfON/In$_{0.53}$Ga$_{0.47}$As MOS capacitors; insets are the hysteresis at 1 MHz.

devices [54]; however, no effort was made to improve the nitridation technology regarding the discussed issues.

As a result, in order to improve the film quality of HfO_2, a cyclic nitrogen low-power plasma step was added within the ALD cycles to passivate oxygen vacancies uniformly without causing damage or surface degradation. Through this technology with a combination of IPA oxidant, achievement of improvement in both interface and bulk quality of high-k/InGaAs MOS properties is expected. The detailed information of the ALD sequence is depicted in **Figure 5**. Every cycle consisted of sequential precursor pulse steps and a gas stabilization step followed by 5 s of 50 W N_2 plasma step with purge steps between pulse steps. It is discovered that there is a trade-off relationship of plasma condition. The plasma power should be enough for effective passivation although it may degrade the substrate by radiation damage. Through the developed ALD sequence, with adequate plasma condition, the HfO_2 layer is improved without having influence in the substrate.

By using the proposed nitridation technology, HfO_xN_y/InGaAs MOS capacitors are fabricated showing promising results as shown in **Figure 6** [41]. A significant suppression of the frequency dispersion was observed upon nitrogen incorporation in every gate bias range. The inversion humps and flat band voltage shift were effectively reduced for all samples, which imply that the defective interface states

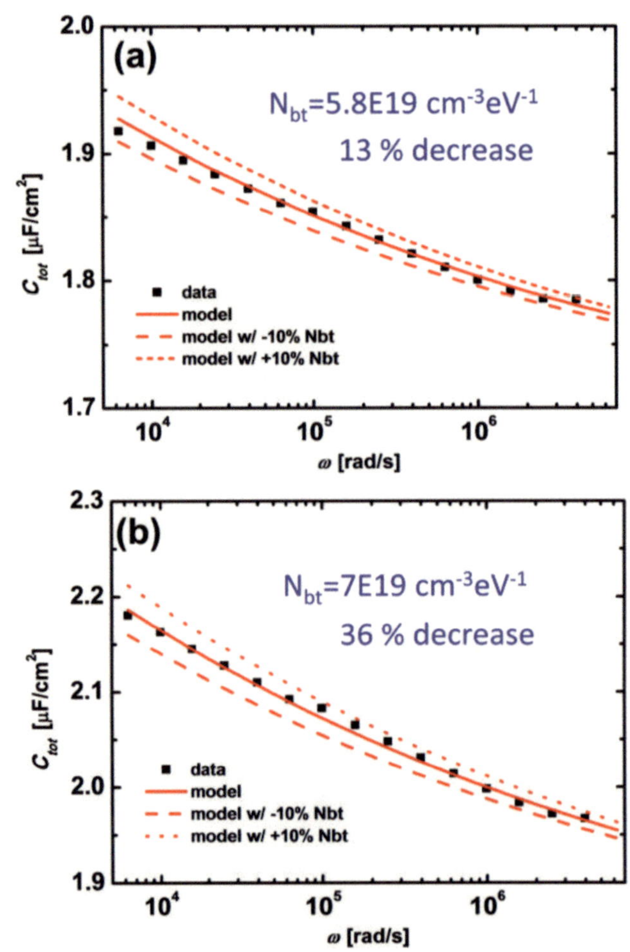

Figure 7.
Border trap estimation of HfON/InGaAs MOS capacitors by using the distributed oxide bulk trap model (a) O_3-HfON and (b) IPA-HfON.

near the mid-gap level can be treated with nitrogen incorporation. It is hypothesized that oxygen diffusion through the oxygen vacancies of HfO_2, which results in the formation of As-Ga anti-sites, was greatly reduced, as oxygen vacancies were effectively passivated with nitrogen [17]. Therefore, nitrogen may block further oxygen diffusion, thereby preventing surface oxidation, which could occur not only during but also after dielectric deposition. Furthermore, the frequency dispersion in the accumulation region greatly reduced to 2.1 and 3.2% per decade for O_3- and IPA-based ALD, respectively. These values are comparable to suppressed dispersion values in low-EOT gate stacks, which imply excellent reliability quality of dielectric stacks on III-V substrate [55]. As the proposed nitridation technology is aimed to treat inferior bulk qualities of HfO_2, it showed greater impact on IPA-based HfO_2. The inversion behavior was observed for the IPA-based ALD HfO_2 which has not been reported from the former HfO_2/InGaAs MOS studies, and it will be further discussed (**Figure 7**).

3. Characterization of IPA-based ALD HfO_2 on n- and p-type InGaAs substrates

Based on the n-type MOS results, the D_{it} was extracted using the conductance method as shown in **Figure 8**. The combination of IPA oxidant and PA-ALD HfO_xN_y with standard interface treatments resulted in a reduced D_{it} level of 4.5×1011 eV^{-1} cm^{-2} at $E_c - E_t = 0.3$ eV. Based on the D_{it} distribution, it is evident that the inversion behavior observed in the C-V curves is due to the significant mid-gap D_{it} decrease, which is consistent with the previously reported studies. As the mid-gap D_{it} is known to correlate to the As-Ga anti-site defect and the ozone-based HfO_xN_y lacked inversion characteristics with high mid-gap D_{it}, these defects might be the reason why the inversion behavior is difficult to be achieved in most studies. Also, these defects might be formed in the initial ALD steps through the oxidant exposure. Also, in **Figure 9**, we have benchmarked our results, comparing them

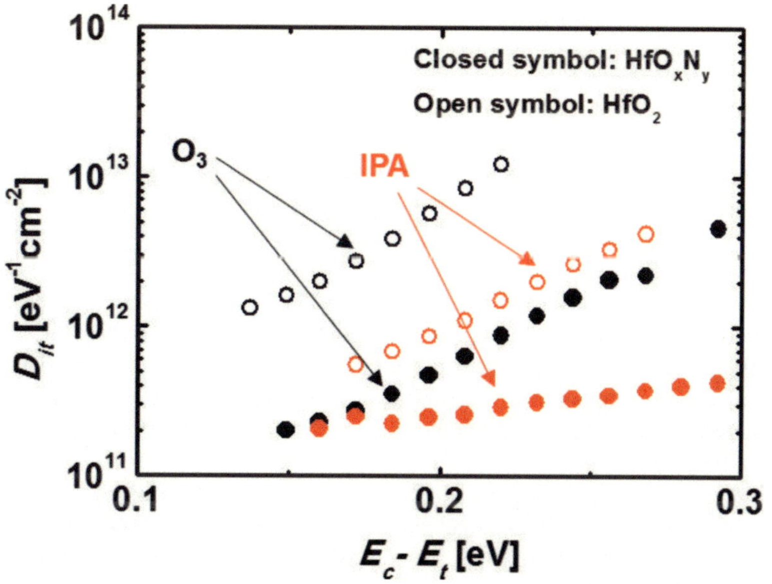

Figure 8.
The D_{it} distribution of the fabricated III-V MOS capacitors showing great reduction in the mid-gap D_{it} with the IPA-based HfO_xN_y.

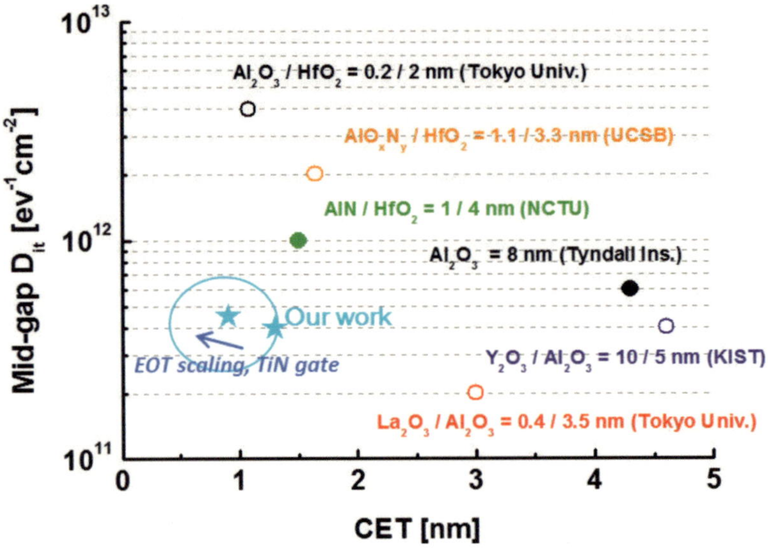

Figure 9.
Benchmarking the mid-gap D_{it} values of the proposed high-k ALD compared to the best results in the field of III-V MOS. The filled circles represent C-V curves with inversion behavior.

to the best results ever reported in the field of III-V MOS device studies [56–61]. Extraordinary mid-gap D_{it} values are achieved with low CET values with the proposed technology. Especially, while other studies mostly suffer from insufficient dielectric constant of the IL, our work employs HfO_2 as an IL, which has merit in terms of the EOT scaling.

In addition, with conductance method in the measurement frequency range of 1 kHz–1 MHz, the n-type MOS capacitor results can only provide information of D_{it} distribution near the conduction band. In order to estimate the total D_{it} distribution throughout the bandgap in InGaAs, p-type InGaAs MOS was fabricated and analyzed. The p-type InGaAs MOS capacitors are fabricated in the same process flow of n-type MOS capacitors.

The multifrequency C-V measurements of IPA-based PA-ALD HfO_xN_y/n-$In_{0.53}Ga_{0.47}As$ MOS capacitors compared to the HfO_2 (O_3) sample is presented in **Figure 10**. Compared to the reference, the optimized HfON process exhibits significant frequency dispersion suppression with steeper C-V slope. This result is comparable to the previously reported high-quality p-type InGaAs MOS results. It is assumed that the interface improvement mechanism is similar to the previous n-type MOS analysis.

Based on the results, the D_{it} distribution within the InGaAs bandgap is extracted with the conductance method shown in **Figure 11**. The D_{it} level at the exact mid-gap energy level ($E_g/2$ = 0.375 eV) is around 8×10^{11} eV^{-1} cm^{-2}. This value is still low for reported III-V MOS interface, and it is suggested that based on the D_{it} distribution, the accumulation mode n-channel III-V devices are favorable than the inversion mode p-substrate III-V devices because the overall D_{it} levels are much lower at the conduction band area.

Moreover, temperature-dependent conductance method was performed in order to analyze the mid-gap D_{it} level thoroughly. High-temperature (350, 400 and 450 K) multifrequency C-V analysis was conducted on HfON/InGaAs MOS capacitors. The C-V results of each measurement temperature are shown in **Figure 12**.

As the measurement temperature increases, the inversion response gets stronger at higher frequencies compared to the room temperature-measured results. Also,

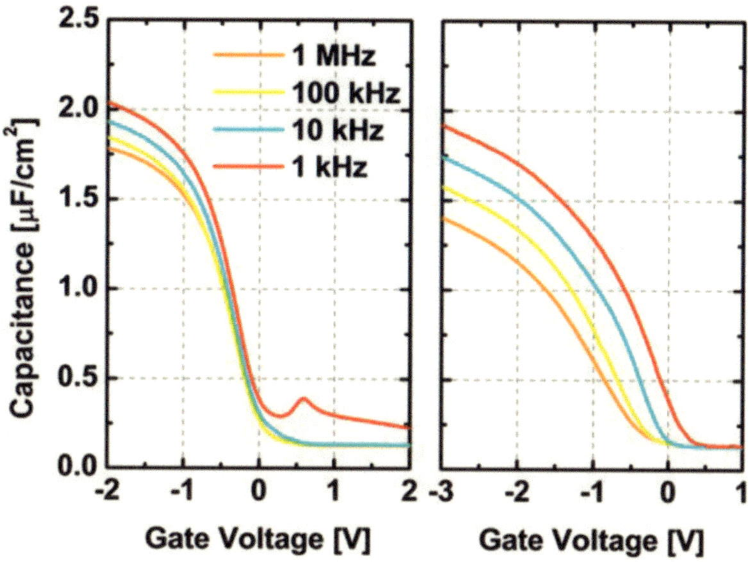

Figure 10.
C-V characteristics of IPA-based PA-ALD HfO_xN_y (left) and O_3-based HfO_2 (right) on p-type $In_{0.53}Ga_{0.47}As$ substrates.

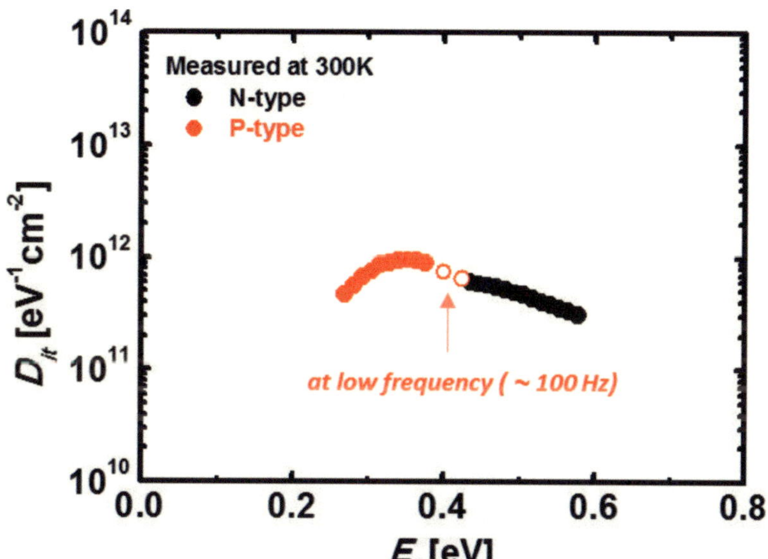

Figure 11.
The total D_{it} distribution within the $In_{0.53}Ga_{0.47}As$ bandgap, which is extracted from the n- and p-type MOS capacitors.

while the dispersion at the accumulation region seems to be similar, there was a significant impact on the inversion hump phenomenon which is the interface trap characteristic. The measurement noise at higher temperature and lower frequencies was also noted. When the measurement temperature reaches around 450 K, the strong inversion response occurs even at 1 MHz, and it interferes with the interface trap-related conductance peak making the deconvolution process impossible.

The D_{it} distribution was estimated from the temperature-dependent conductance technique as shown in **Figure 13**. As the measurement temperature increases, the

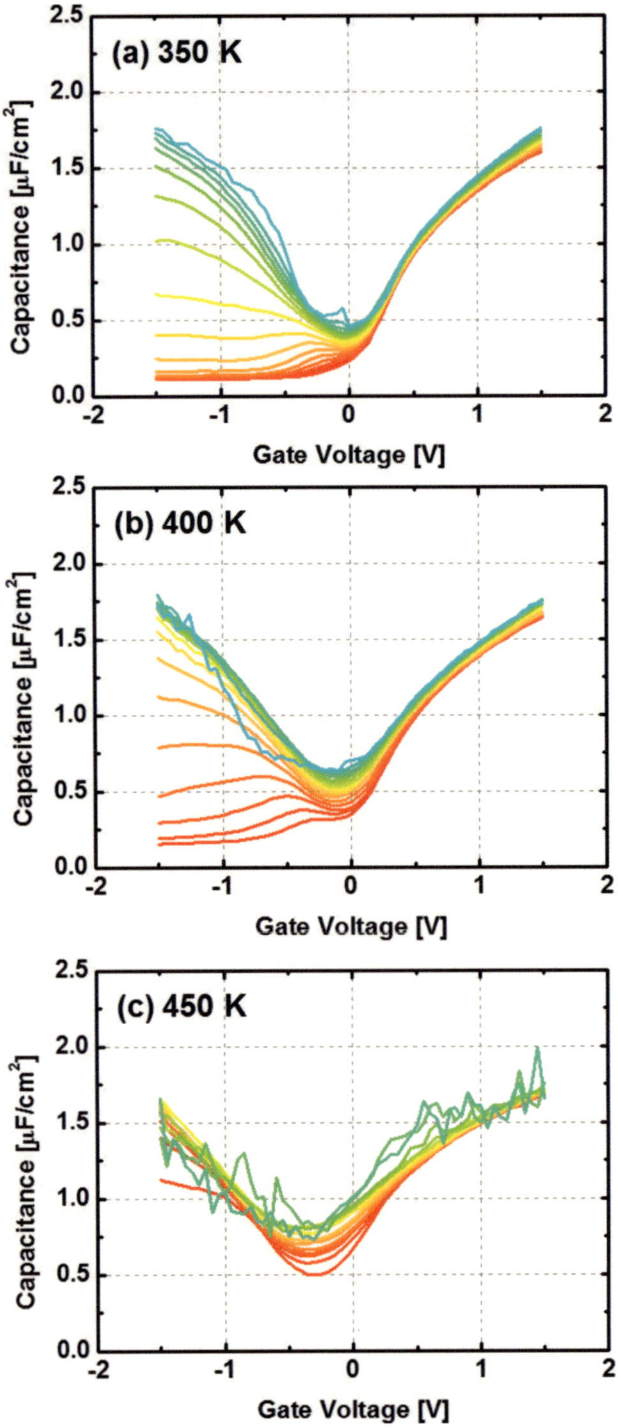

Figure 12.
C-V characteristics of IPA-based PA-ALD HfO_xN_y on n-type InGaAs substrate measured at (a) 350 K, (b) 400 K, and (c) 450 K.

deeper energy range could be measured. Similar D_{it} profile was observed showing a peak energy level around the exact mid-gap level (~0.375 eV). The peak D_{it} value is slightly higher than the previously estimated value which would be the effect of

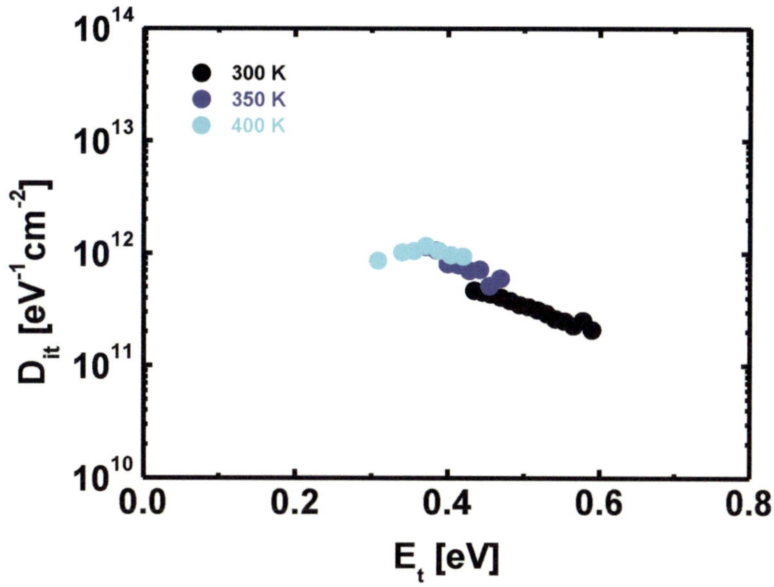

Figure 13.
The total D_{it} distribution within the $In_{0.53}Ga_{0.47}As$ bandgap, which is extracted from the temperature-dependent conductance method.

enhanced thermal broadening of trap response in higher temperatures. The differences between the D_{it} profile estimation can be summarized as follows: While using p-MOS capacitors, a larger energy level range is observable with no thermal broadening of the trap response due to the fixed measurement temperature. On the other hand, using a temperature-dependent method has a thermal broadening issue but only requires one sample for characterization.

It was noted that the inversion behavior of IPA-based HfO_xN_y ALD is attributed to the mid-gap D_{it} level decrease. However, in order to verify true inversion characteristics, more analyses must be investigated. In **Figure 14**, the conductance profile of sample O_3-based HfO_2 and IPA-based HfO_xN_y InGaAs MOS capacitors are depicted.

Both C-V profiles have shown inversion-like behavior in the negative bias region. However, clear difference is observed between the conductance profiles. While huge and Gaussian conductance profiles in the negative bias regions are observed for O_3-based HfO_2, smaller Gaussian conductance peaks are observed in depletion region, and distinct from these peaks, saturated conductance profiles are observed for IPA-based HfO_xN_y. Therefore, it is concluded that the inversion like behavior in O_3-based HfO_2 is attributed from the huge and broad conductance peaks that reflect high mid-gap D_{it} levels, while inversion behavior in IPA-based HfO_xN_y might be attributed from real true minority carrier inversion.

In **Figure 15**, to verify true inversion characteristics of IPA-based HfO_xN_y, the minority carrier response was investigated based on the extraction of the transition frequency, w_m, which is known to be a characteristic of a strong inverted surface for III-V MOS capacitors [62]. It is known that at the transition frequency, the $-wdC/dw$ and G_m/w share the same peak magnitude in the strong inversion gate bias. Notably, $-wdC/dw$ and G_m/w share the same peak magnitude at the same transition frequency of 4 kHz which suggests that IPA-based HfO_xN_y exhibits true inversion behavior. The true inversion behavior of hafnium oxide-based dielectrics on InGaAs substrate has not been reported yet which implies significant potential.

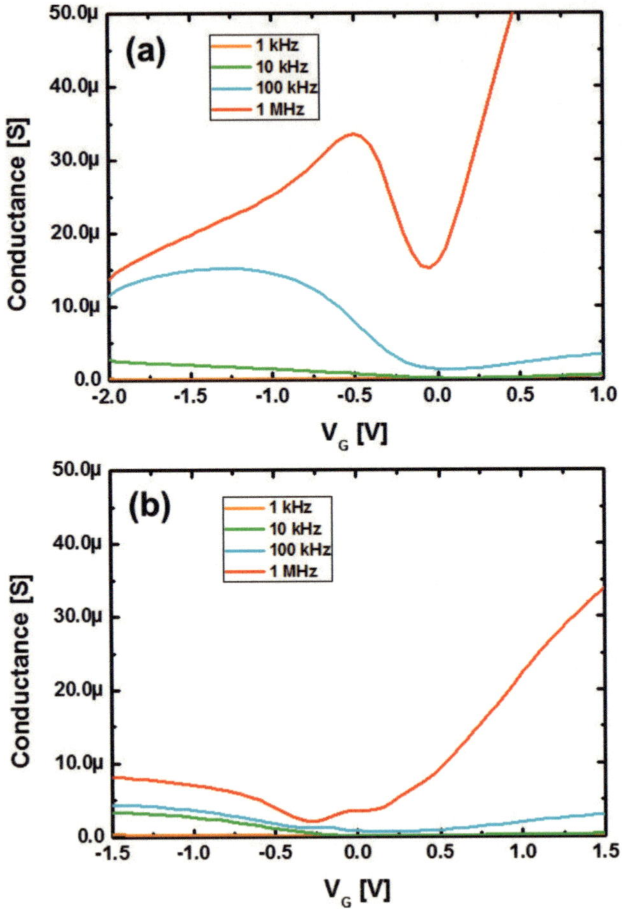

Figure 14.
The conductance profiles for (a) O_3-HfO_2 and (b) IPA-HfON ALD.

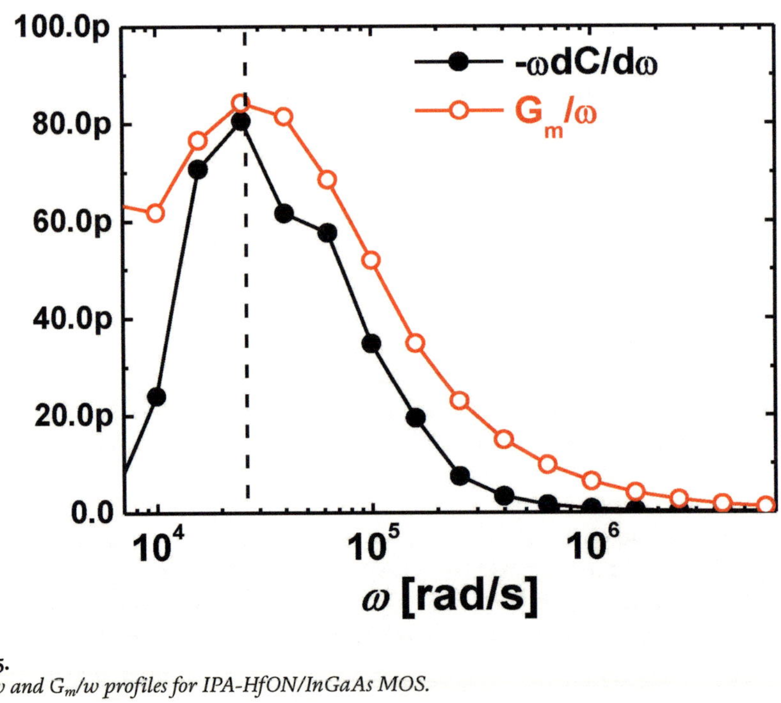

Figure 15.
$-wdC/dw$ and G_m/w profiles for IPA-HfON/InGaAs MOS.

4. Conclusions

In order to achieve both low EOT and low D_{it}, a highly advanced gate stack, prepared by using an IPA oxidant in the PA-ALD of HfO_xN_y on $In_{0.53}Ga_{0.47}As$ substrates, was proposed and showed the most outstanding results. A cyclic nitrogen low-power plasma step was added within the ALD cycles to passivate the oxygen vacancies uniformly without causing damage or surface degradation in comparison to the post deposition nitridation technology. Remarkable mid-gap D_{it} levels with strong inversion characteristics were achieved which has not been reported in the previous HfO_2/InGaAs interface studies. The improved interface characteristics can be attributed to both low surface oxidation ability of IPA and suppression of oxygen diffusion by effective nitrogen passivation to oxygen vacancies in HfO_2. The proposed ALD HfO_xN_y was fully characterized by investigating different dopant types and measurement temperatures. The results show comprehensive understanding on the interface defect density distribution. It is suggested that not only surface treatments but also the development of an advanced HfO_2 ALD process has a great impact on the quality of the III-V MOS interface and the IPA-based HfON interfacial layer might have great potential in future technology node.

Author details

Sukeun Eom*, Min-woo Kong and Kwang-seok Seo
Seoul National University, Seoul, Republic of Korea

*Address all correspondence to: djatnrms90@snu.ac.kr

IntechOpen

© 2020 The Author(s). Licensee IntechOpen. This chapter is distributed under the terms of the Creative Commons Attribution License (http://creativecommons.org/licenses/by/3.0), which permits unrestricted use, distribution, and reproduction in any medium, provided the original work is properly cited. CC BY

References

[1] Cho H-J, Oh HS, Nam KJ, Kim YH, Yeo KH, Kim WD, et al. Si FinFET based 10nm technology with multi Vt gate stack for low power and high performance applications. In: 2016 IEEE Symposium on VLSI Technology. IEEE; 2016. DOI: 10.1109/VLSIT.2016.7573359

[2] Auth C, Aliyarukunju A, Asoro M, Bergstrom D, Bhagwat V, Birdsall J, et al. A 10nm high performance and low-power CMOS technology featuring 3rd generation FinFET transistors, Self-Aligned Quad Patterning, contact over active gate and cobalt local interconnects. In: 2017 IEEE International Electron Devices Meeting (IEDM). IEEE; 2017. DOI: 10.1109/IEDM.2017.8268472

[3] Tsai T-H, Sheen R-B, Chang C-H, Staszewski RB. A 0.2GHz to 4GHz Hybrid PLL (ADPLL/Charge-Pump-PLL) in 7NM FinFET CMOS Featuring 0.619PS Integrated Jitter and 0.6US Settling Time at 2.3MW. In: 2018 IEEE Symposium on VLSI Circuits. IEEE; 2018. DOI: 10.1109/VLSIC.2018.8502274

[4] Jeong WC, Kwon DJ, Nam KJ, Rim WJ, Jang MS, Kim HT, et al. True 7nm Platform Technology featuring Smallest FinFET and Smallest SRAM cell by EUV, Special Constructs and 3rd Generation Single Diffusion Break. In: 2018 IEEE Symposium on VLSI Technology. IEEE; 2018. DOI: 10.1109/VLSIT.2018.8510682

[5] Bohr MT, Young IA. CMOS scaling trends and beyond. IEEE Micro. 2017;**37**(6):20-29. DOI: 10.1109/MM.2017.4241347

[6] Theis TN, Wong H-SP. The end of Moore's Law: A new beginning for information technology. Computing in Science & Engineering. 2017;**19**(2): 41-50. DOI: 10.1109/MCSE.2017.29

[7] Duh KHG, Chao PC, Liu SMJ, Ho P, Kao MY, Ballingall JM. A super low-noise 0.1 mu m T-gate InAlAs-InGaAs-InP HEMT. IEEE Microwave and Guided Wave Letters. 1991;**1**(5): 114-116. DOI: 10.1109/75.89081

[8] Lai R, Mei XB, Deal WR, Yoshida W, Kim YM, Liu PH, et al. Sub 50 nm InP HEMT Device with Fmax Greater than 1 THz. In: 2007 IEEE International Electron Devices Meeting. IEEE; 2007. DOI: 10.1109/IEDM.2007.4419013

[9] Lin J, Antoniadis DA, del Alamo JA. Sub-30 nm InAs Quantum-Well MOSFETs with self-aligned metal contacts and Sub-1 nm EOT HfO_2 insulator. In: 2012 International Electron Devices Meeting. IEEE; 2012. DOI: 10.1109/IEDM.2012.6479149

[10] Kim D-H, del Alamo JA. Logic Performance of 40 nm InAs HEMTs. In: 2007 IEEE International Electron Devices Meeting. IEEE; 2007. DOI: 10.1109/IEDM.2007.4419018

[11] Yan R-H, Ourmazd A, Lee KF. Scaling the Si MOSFET: From bulk to SOI to bulk. IEEE Transactions on Electron Devices. 1992 Jul;**39**(7): 1704-1710. DOI: 10.1109/16.141237

[12] Cirlin GE, Dubrovskii VG, Soshnikov IP, Sibirev NV, Samsonenko YB, Bouravleuv AD, et al. Critical diameters and temperature domains for MBE growth of III-V nanowires on lattice mismatched substrates. Physica Status Solidi RRL: Rapid Research Letters. 2009;**3**(4): 112-114. DOI: 10.1002/pssr.200903057

[13] Daix N, Uccelli E, Czornomaz L, Caimi D, Rossel C, Sousa M, et al. Towards large size substrates for III-V co-integration made by direct wafer bonding on Si. APL Materials. 2014 Aug;**2**(8):086104. DOI: 10.1063/1.4893653

[14] Czornomaz L, Daix N, Caimi D, Sousa M, Erni R, Rossell MD, et al. An

integration path for gate-first UTB III-V-on-insulator MOSFETs with silicon, using direct wafer bonding and donor wafer recycling. In: 2012 International Electron Devices Meeting. IEEE; 2012. DOI: 10.1109/IEDM.2012.6479088

[15] Li JZ, Bai J, Park J-S, Adekore B, Fox K, Carroll M, et al. Defect reduction of GaAs epitaxy on Si (001) using selective aspect ratio trapping. Applied Physics Letters. 2007;**91**(2):021114. DOI: 10.1063/1.2756165

[16] Waldron N, Merckling C, Guo W, Ong P, Teugels L, Ansar S, et al. An InGaAs/InP quantum well finfet using the replacement fin process integrated in an RMG flow on 300mm Si substrates. In: 2014 Symposium on VLSI Technology (VLSI-Technology): Digest of Technical Papers. IEEE; 2014. DOI: 10.1109/VLSIT.2014.6894349

[17] Robertson J, Guo Y, Lin L. Defect state passivation at III-V oxide interfaces for complementary metal–oxide–semiconductor devices. Journal of Applied Physics. 2015;**117**(11):112806. DOI: 10.1063/1.4913832

[18] Brammertz G, Lin HC, Martens K, Alian A-R, Merckling C, Penaud J, et al. Electrical properties of III-V/oxide interfaces. ECS Transactions. 2009;**19**(5):375-386. DOI: 10.1149/1.3119560

[19] Hinkle CL, Milojevic M, Brennan B, Sonnet AM, Aguirre-Tostado FS, Hughes GJ, et al. Detection of Ga suboxides and their impact on III-V passivation and Fermi-level pinning. Applied Physics Letters. 2009;**94**(16):162101. DOI: 10.1063/1.3120546

[20] Shahrjerdi D, Tutuc E, Banerjee SK. Impact of surface chemical treatment on capacitance-voltage characteristics of GaAs metal-oxide-semiconductor capacitors with Al_2O_3 gate dielectric. Applied Physics Letters. 2007;**91**(6):063501. DOI: 10.1063/1.2764438

[21] He G, Chen X, Sun Z. Interface engineering and chemistry of Hf-based high-k dielectrics on III–V substrates. Surface Science Reports. 2013;**68**(1):68-107. DOI: 10.1016/j.surfrep.2013.01.002

[22] Hollinger G, Skheyta-Kabbani R, Gendry M. Oxides on GaAs and InAs surfaces: An x-ray-photoelectron-spectroscopy study of reference compounds and thin oxide layers. Physical Review B. 1994;**49**(16):11159-11167. DOI: 10.1103/PhysRevB.49.11159

[23] Ye PD, Wilk GD, Kwo J, Yang B, Gossmann H-JL, Frei M, et al. GaAs MOSFET with oxide gate dielectric grown by atomic layer deposition. IEEE Electron Device Letters. 2003;**24**(4):209-211. DOI: 10.1109/LED.2003.812144

[24] Huang YL, Chang P, Yang ZK, Lee YJ, Lee HY, Liu HJ, et al. Thermodynamic stability of Ga2O3(Gd2O3)/GaAs interface. Applied Physics Letters. 2005;**86**(19):191905. DOI: 10.1063/1.1923172

[25] Poindexter EH, Gerardi GJ, Rueckel ME, Caplan PJ, Johnson NM, Biegelsen DK. Electronic traps and Pb centers at the Si/SiO2interface: Band-gap energy distribution. Journal of Applied Physics. 1984;**56**(10): 2844-2849. DOI: 10.1063/1.333819

[26] Engel-Herbert R, Hwang Y, Stemmer S. Quantification of trap densities at dielectric/III–V semiconductor interfaces. Applied Physics Letters. 2010;**97**(6):062905. DOI: 10.1063/1.3479047

[27] Stemmer S, Chobpattana V, Rajan S. Frequency dispersion in III-V metal-oxide-semiconductor capacitors. Applied Physics Letters. 2012;**100**(23):233510. DOI: 10.1063/1.4724330

[28] Engel-Herbert R, Hwang Y, Stemmer S. Comparison of methods to quantify interface trap densities at dielectric/III-V semiconductor interfaces. Journal of Applied Physics. 2010;**108**(12):124101. DOI: 10.1063/1.3520431

[29] Yuan Y, Yu B, Ahn J, McIntyre PC, Asbeck PM, Rodwell MJW, et al. A distributed bulk-oxide trap model for Al_2O_3 InGaAs MOS devices. IEEE Transactions on Electron Devices. 2012;**59**(8):2100-2106. DOI: 10.1109/TED.2012.2197000

[30] Chobpattana V, Son J, Law JJM, Engel-Herbert R, Huang C-Y, Stemmer S. Nitrogen-passivated dielectric/InGaAs interfaces with sub-nm equivalent oxide thickness and low interface trap densities. Applied Physics Letters. 2013;**102**(2):022907. DOI: 10.1063/1.4776656

[31] Kent T, Tang K, Chobpattana V, Negara MA, Edmonds M, Mitchell W, et al. The influence of surface preparation on low temperature HfO_2 ALD on InGaAs (001) and (110) surfaces. The Journal of Chemical Physics. 2015;**143**(16):164711. DOI: 10.1063/1.4934656

[32] Rafí JM, Zabala M, Beldarrain O, Campabadal F. Effect of processing conditions on the electrical characteristics of atomic layer deposited Al_2O_3 and HfO_2 films. ECS Transactions. 2010;**28**(2):213. DOI: 10.1149/1.3372577

[33] García H, Castán H, Dueñas S, Bailón L, Campabadal F, Beldarrain O, et al. Electrical characterization of atomic-layer-deposited hafnium oxide films from hafnium tetrakis(dimethylamide) and water/ozone: Effects of growth temperature, oxygen source, and postdeposition annealing. Journal of Vacuum Science & Technology, A: Vacuum, Surfaces, and Films. 2013;**31**(1):01A127. DOI: 10.1116/1.4768167

[34] Ahn J, McIntyre PC. Subcutaneous oxidation of $In_{0.53}Ga_{0.47}As(100)$ through ultra-thin atomic layer deposited Al_2O_3. Applied Physics Letters. 2013;**103**(25):251602. DOI: 10.1063/1.4850520

[35] Cheng C-W, Fitzgerald EA. In situ metal-organic chemical vapor deposition atomic-layer deposition of aluminum oxide on GaAs using trimethyaluminum and isopropanol precursors. Applied Physics Letters. 2008;**93**(3):031902. DOI: 10.1063/1.2960574

[36] Parks GS, Barton B. Vapor pressure data for isopropyl alcohol and tertiary butyl alcohol. Journal of the American Chemical Society. 1928;**50**(1):24-26. DOI: 10.1021/ja01388a004

[37] Swerts J, Peys N, Nyns L, Delabie A, Franquet A, Maes JW, et al. Impact of precursor chemistry and process conditions on the scalability of ALD HfO[sub 2] gate dielectrics. Journal of the Electrochemical Society. 2010;**157**(1):G26. DOI: 10.1149/1.3258664

[38] Krylov I, Eizenberg M, Ritter D. Determination of the dielectric constant of InGaAs based gate stacks by a modified thickness series method. Applied Physics Letters. 2014;**105**(20):203506. DOI: 10.1063/1.4902114

[39] Harrell WR, Frey J. Observation of Poole–Frenkel effect saturation in SiO2 and other insulating films. Thin Solid Films. 1999;**352**(1-2):195-204. DOI: 10.1016/S0040-6090(99)00344-2

[40] Simoen E, Lin DH-C, Alian A, Brammertz G, Merckling C, Mitard J, et al. Border traps in Ge/III–V channel devices: Analysis and reliability aspects. IEEE Transactions on Device and Materials Reliability. 2013;**13**(4):444-455. DOI: 10.1109/TDMR.2013.2275917

[41] Eom S-K, Kong M-W, Kang M-J, Lee J-G, Cha H-Y, Seo K-S. Enhanced interface characteristics of PA-ALD HfOxNy/InGaAs MOSCAPs using IPA oxygen reactant and cyclic N2 plasma. IEEE Electron Device Letters. 2018;**39**(11):1636-1639. DOI: 10.1109/LED.2018.2870176

[42] Gavartin JL, Muñoz Ramo D, Shluger AL, Bersuker G, Lee BH. Negative oxygen vacancies in HfO_2 as charge traps in high-k stacks. Applied Physics Letters. 2006;**89**(8):082908. DOI: 10.1063/1.2236466

[43] Greeuw G, Verwey JF. The mobility of Na+, Li+, and K+ ions in thermally grown SiO2 films. Journal of Applied Physics. 1984;**56**(8):2218-2224. DOI: 10.1063/1.334256

[44] Trivedi AR, Ando T, Singhee A, Kerber P, Acar E, Frank DJ, et al. A simulation study of oxygen vacancy-induced variability in HfO_2/metal gated SOI FinFET. IEEE Transactions on Electron Devices. 2014;**61**(5):1262-1269. DOI: 10.1109/TED.2014.2313086

[45] Xiong K, Robertson J, Clark SJ. Passivation of oxygen vacancy states in HfO_2 by nitrogen. Journal of Applied Physics. 2006;**99**(4):044105. DOI: 10.1063/1.2173688

[46] Lee BH, Kang L, Nieh R, Qi W-J, Lee JC. Thermal stability and electrical characteristics of ultrathin hafnium oxide gate dielectric reoxidized with rapid thermal annealing. Applied Physics Letters. 2000;**76**(14):1926-1928. DOI: 10.1063/1.126214

[47] Dalapati GK, Sridhara A, Wong ASW, Chia CK, Chi DZ. HfOxNy gate dielectric on p-GaAs. Applied Physics Letters. 2009;**94**(7):073502. DOI: 10.1063/1.3079409

[48] Jin CG, Yang Y, Zhang HY, Huang TY, Wu MZ, Zhuge LJ, et al. Controllable nitrogen incorporation in HfO_2 films by modulating capacitively-combined inductively-coupled plasmas. Journal of Physics D: Applied Physics. 2013;**46**(48):485206. DOI: 10.1088/0022-3727/46/48/485206

[49] Lee YB, Oh I-K, Cho EN, Moon P, Kim H, Yun I. Characterization of HfO N thin film formation by in-situ plasma enhanced atomic layer deposition using NH3 and N2 plasmas. Applied Surface Science. 2015;**349**:757-762. DOI: 10.1016/j.apsusc.2015.05.066

[50] Kang CS, Cho H-J, Choi R, Kim YH, Kang CY, Rhee SJ, et al. The electrical and material characterization of hafnium oxynitride gate dielectrics with TaN-gate electrode. IEEE Transactions on Electron Devices. 2004;**51**(2):220-227. DOI: 10.1109/TED.2003.821707

[51] Yu X, Zhu C, Yu M. Impact of nitrogen in HfON gate dielectric with metal gate on electrical characteristics, with particular attention to threshold voltage instability. Applied Physics Letters. 2007;**90**(10):103502. DOI: 10.1063/1.2709948

[52] Dai M, Wang Y, Shepard J, Liu J, Brodsky M, Siddiqui S, et al. Effect of plasma N2 and thermal NH3 nitridation in HfO_2 for ultrathin equivalent oxide thickness. Journal of Applied Physics. 2013;**113**(4):044103. DOI: 10.1063/1.4775817

[53] Cheng C C, Chien C H, Chen C W, Hsu S-L, Yang M-Y, Huang C-C, et al. Impact of post-deposition-annealing on the electrical characteristics of HfOxNy gate dielectric on Ge substrate. Microelectronic Engineering. 2005;**80**:30-33. DOI: 10.1016/j.mee.2005.04.033

[54] Chang P-C, Luc Q-H, Lin Y-C, Lin Y-K, Wu C-H, Sze SM, et al. InGaAs QW-MOSFET performance improvement using a PEALD-AlN passivation layer and an *In-situ* NH3Post

remote-plasma treatment. IEEE Electron Device Letters. 2017;**38**(3): 310-313. DOI: 10.1109/LED. 2017.2656180

[55] Vais A, Franco J, Martens K, Lin D, Sioncke S, Putcha V, et al. A new quality metric for III–V/high-k MOS gate stacks based on the frequency dispersion of accumulation capacitance and the CET. IEEE Electron Device Letters. 2017;**38**(3):318-321. DOI: 10.1109/ LED.2017.2657794

[56] Hoshii T, Lee S, Suzuki R, Taoka N, Yokoyama M, Yamada H, et al. Reduction in interface state density of Al_2O_3/InGaAs metal-oxide-semiconductor interfaces by InGaAs surface nitridation. Journal of Applied Physics. 2012;**112**(7):073702. DOI: 10.1063/1.4755804

[57] Suzuki R, Taoka N, Yokoyama M, Lee S, Kim SH, Hoshii T, et al. 1-nm-capacitance-equivalent-thickness HfO_2/Al_2O_3/InGaAs metal-oxide-semiconductor structure with low interface trap density and low gate leakage current density. Applied Physics Letters. 2012;**100**(13):132906. DOI: 10.1063/1.3698095

[58] Lee S, Chobpattana V, Huang C-Y, Thibeault BJ, Mitchell W, Stemmer S, et al. Record Ion (0.50 mA/um at VDD = 0.5 V and Ioff = 100 nA/um) 25 nm-gate-length ZrO2/InAs/InAlAs MOSFETs. In: 2014 Symposium on VLSI Technology (VLSI-Technology): Digest of Technical Papers. IEEE; 2014. DOI: 10.1109/VLSIT.2014.6894363

[59] Chang C-Y, Ichikawa O, Osada T, Hata M, Yamada H, Takenaka M, et al. Impact of La_2O_3 interfacial layers on InGaAs metal-oxide-semiconductor interface properties in Al_2O_3/La_2O_3/ InGaAs gate stacks deposited by atomic-layer-deposition. Journal of Applied Physics. 2015;**118**(8):085309. DOI: 10.1063/1.4929650

[60] Luc Q, Cheng S, Chang P, Do H, Chen J, Ha M, et al. Effects of in-situ plasma-enhanced atomic layer deposition treatment on the performance of $HfO_2/In_{0.53}Ga_{0.47}As$ metal-oxide-semiconductor field-effect transistors. IEEE Electron Device Letters. 2016;**1**:1. DOI: 10.1109/ LED.2016.2581175

[61] Kim SK, Geum D-M, Shim J-P, Kim CZ, Kim H, Song JD, et al. Fabrication and characterization of Pt/ $Al_2O_3/Y_2O_3/In_{0.53}Ga_{0.47}As$ MOSFETs with low interface trap density. Applied Physics Letters. 2017;**110**(4):043501. DOI: 10.1063/1.4974893

[62] O'Connor É, Cherkaoui K, Monaghan S, Sheehan B, Povey IM, Hurley PK. Effect of forming gas annealing on the inversion response and minority carrier generation lifetime of n and p-$In_{0.53}Ga_{0.47}As$ MOS capacitors. Microelectronic Engineering. 2015;**147**:325-329. DOI: 10.1016/j. mee.2015.04.103

Chapter 8

Surface-Enhanced Raman Scattering: Introduction and Applications

Samir Kumar, Prabhat Kumar, Anamika Das and Chandra Shakher Pathak

Abstract

Scattering of light by molecules can be elastic, Rayleigh scattering, or inelastic, Raman scattering. In the elastic scattering, the photon's energy and the state of the molecule after the scattering events are unchanged. Hence, Rayleigh scattered light does not contain much information on the structure of molecular states. In inelastic scattering, the frequency of monochromatic light changes upon interaction with the vibrational states, or modes, of a molecule. With the advancement in the laser sources, better and compact spectrometers, detectors, and optics Raman spectroscopy have developed as a highly sensitive technique to probe structural details of a complex molecular structure. However, the low scattering cross section (10^{-31}) of Raman scattering has limited the applications of the conventional Raman spectroscopy. With the discovery of surface-enhanced Raman scattering (SERS) in 1973 by Martin Fleischmann, the interest of the research community in Raman spectroscopy as an analytical method has been revived. This chapter aims to familiarize the readers with the basics of Raman scattering phenomenon and SERS. This chapter will also discuss the latest developments in the SERS and its applications in various fields.

Keywords: Raman scattering, surface-enhanced Raman spectroscopy, enhancement factor, nanoparticles, 2D materials

1. Introduction: Raman scattering

Scattering of light by molecules can be elastic, *Rayleigh scattering*, or inelastic, *Raman scattering*. In the elastic case, the photon's energy and the state of the molecule after the scattering events are unchanged. Hence, Rayleigh scattered light does not contain much information on the structure of molecular states [1]. In inelastic scattering, the frequency of photons of monochromatic light changes upon interaction with the vibrational states, or modes, of a molecule. The effect was postulated theoretically by Smekal et al. in 1923 but was first discovered experimentally by C.V. Raman in 1928 in an experiment using the sun as a light source [2–4].
In Raman scattering, two inelastic processes can occur:

- *Stokes* process: An incident photon $h\nu_0$ excites a molecular vibration $h\nu_{vib}$ and is thus scattered with the corresponding difference in energy $h(\nu_0 - \nu_{vib})$ (red shift).

- *Anti-Stokes* process: The photon acquires vibrational energy and is scattered with a higher energy $h(\nu_0 + \nu_{vib})$ (blue shift).

This shift provides information about vibrational, rotational, and other low-frequency transitions in molecules. Raman spectroscopy can be used to study solid, liquid, and gaseous samples.

1.1 Classical theory of the Raman effect

Raman scattering can be explained using the molecular polarizability [5]. If a molecule is placed in an electric field, electrons and nuclei get displaced. Due to the separation of charged species, an electric dipole moment is induced in the molecule, and it is said to be polarized. If E is the strength of the electric field and μ is the magnitude of the induced dipole moment, then

$$\mu = \alpha E \tag{1}$$

where α is the polarizability of the molecule. If a sample is subjected to an electromagnetic wave of frequency ν, the electric field experienced by each molecule of the sample varies as

$$E = E_0 \cos 2\pi \upsilon t \tag{2}$$

where E_0 is the amplitude of the electromagnetic wave. From Eq. (1)

$$\mu = \alpha E_0 \cos 2\pi \upsilon t \tag{3}$$

Thus, Eq. (3) implies that interaction of electromagnetic radiation of frequency ν induces a molecular dipole moment that oscillates and emits radiation of the same frequency, and this is the classical explanation of Rayleigh scattering. However, the ability to perturb the local electron cloud of a molecular structure depends on the relative location of the individual atoms; hence, the polarizability is a function of the instantaneous position of the constituent atoms. So, the polarizability changes with small displacement from equilibrium position (i.e., molecular vibration) and is given by

$$\alpha = \alpha_0 + \left(q - q_{eq}\right)\frac{\partial \alpha}{\partial q} \tag{4}$$

where α_0 is equilibrium polarizability and q_{eq} and q are bond lengths at equilibrium position and any instant, respectively. If a molecule executes simple harmonic motion, the displacement can be represented as

$$q - q_0 = q_{max} \cos 2\pi \upsilon_{vib} t \tag{5}$$

where ν_{vib} is the vibrational frequency of a molecule and q_{max} is the maximum separation distance between atoms relative to their equilibrium position. Substituting Eq. (5) into Eq. (4) gives

$$\alpha = \alpha_0 + \left(\frac{\partial \alpha}{\partial q}\right) q_{max} \cos 2\pi \upsilon_{vib} t \tag{6}$$

Substituting Eq. (6) into Eq. (3) gives

$$\mu = E_0 \cos 2\pi \upsilon t \left[\alpha_0 + \left(\frac{\partial \alpha}{\partial q} \right) q_{max} \cos 2\pi \upsilon_{vib} t \right] \tag{7}$$

or

$$\mu = \alpha_0 E_0 \cos 2\pi \upsilon t + E_0 \left(\frac{\partial \alpha}{\partial q} \right) q_{max} \cos 2\pi \upsilon t \ \cos 2\pi \upsilon_{vib} t \tag{8}$$

or

$$\mu = \alpha_0 E_0 \cos 2\pi \upsilon t + \frac{E_0}{2} q_{max} \left(\frac{\partial \alpha}{\partial q} \right) \cos \left[2\pi (\upsilon - \upsilon_{vib}) t \right] + \frac{E_0}{2} q_{max} \left(\frac{\partial \alpha}{\partial q} \right) \cos \left[2\pi (\upsilon + \upsilon_{vib}) t \right] \tag{9}$$

The first term in Eq. (9) represents Rayleigh scattering and occurs at the excitation frequency ν. The second and third terms correspond to Stokes ($\nu - \nu_{vib}$) and anti-Stokes ($\nu + \nu_{vib}$) scattering. In both inelastic scatterings, the excitation frequency is modulated by the vibrational frequency of the bond.

Besides, from Eq. (9), the molecules that have Raman-active vibration modes must experience a change in polarizability during a vibration $\left(\frac{\partial \alpha}{\partial q}\right)$, i.e., the electron density in the molecule must distort from its typical shape (inducing a dipole). Molecules with symmetrical bends and stretches, therefore, are generally better Raman scatterers. So, for a molecule to be Raman active, its molecular rotation or vibration must cause a change in a component of molecular polarizability.

1.2 Quantum theory of the Raman effect

Raman scattering can be easily understood in terms of the quantum theory of radiation. In the quantum model, the molecules exist in quantized energy levels corresponding to possible stationary states of the molecule. When radiation having an energy $h\nu$ incident on a sample, it is considered that the photons undergo collisions with the molecules. When the collision is elastic, the photons will be

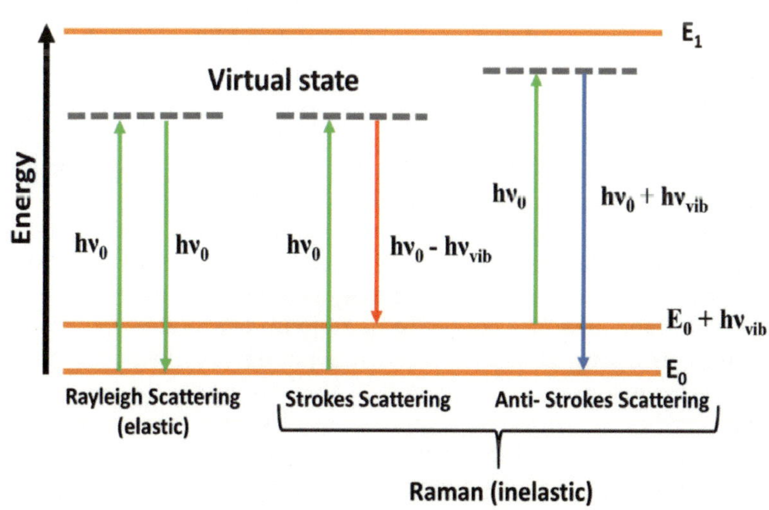

Figure 1.
Energy level diagram for Rayleigh and Raman scattering, where $\Delta E = h\nu_{vib}$ *represents the difference in vibration energy levels [7].*

deflected unchanged, but it is also possible that during the collision, energy is exchanged between the photon and molecule, and as a result, the molecule can gain or lose energy ΔE, where $\Delta E = h\nu_{vib}$ represents a difference in the vibrational or rotational energy levels of that molecule [6]. In quantum mechanical terms, the scattering can be considered as an excitation to a virtual state lower in energy than a real electronic state. When the molecule gains energy ΔE, the photon will be scattered with energy $h\nu - h\nu_{vib}$, and the scattering is known as Stokes' scattering. Conversely, if the molecule loses energy ΔE, the scattered photon will have energy $h\nu + h\nu_{vib}$, and this type of scattering is known as anti-Stokes' scattering. Generally, Stokes' radiation is stronger than the anti-Stokes' radiation. **Figure 1** illustrates the energy level diagram for scattering [7].

2. Surface-enhanced Raman spectroscopy

One of the limitations of the Raman effect is that it is a very weak phenomenon. About one in 10^7 photons undergo Raman scattering. Therefore the Raman signal is very low from low concentrations of the analyte or poor Raman scatterers. Sometimes the high fluorescence from the molecule obscures the Raman signals. Surface-enhanced Raman spectroscopy (SERS) is all about amplifying Raman signals from molecules, by several orders of magnitude [8]. SERS is a technique where molecules undergo much higher scattering efficiencies when adsorbed on metal colloidal nanoparticles or rough metal surfaces. The SERS effect was discovered in 1974 by Fleischmann et al. [9]. The group discovered an anomalously large enhancement of the Raman signal of pyridine in the presence of a roughened silver electrode. The enhancement was initially attributed to greater than expected, or fractal-like, surface area, but subsequent reports showed that the anomalous intensity could not be accounted for by increased surface area and was, in fact, a new phenomenon, giving rise to the idea of the SERS cross-section [10, 11]. However, while SERS has become a large and extremely active field of study, there is still a debate on the exact details of its mechanism and its magnitude [12, 13].

Since then, several enhancement mechanisms were proposed in the early days of SERS. However, only two mechanisms are now broadly accepted, i.e., electromagnetic (EM) theory and chemical enhancement (CE) theory [8, 14, 15]. The electromagnetic models treat the molecule as a point dipole which responds to the enhanced local fields at or near the metal surface [16]. These enhanced fields, in turn, arise from roughness features that couple the incident field to surface plasmons [17]. On the other hand, chemical models attribute SERS intensity to modification of the molecular polarizability by interaction with the metal with ensuing molecular resonances, giving rise to enhancements such as those associated with resonance Raman scattering [18]. CE theory depends on the chemical interaction between probe molecules and the noble metal and is said to contribute only a maximum of about two to three orders of magnitude [19]. Both of these enhancements work simultaneously but are yet to be fully understood because of the difficulties in investigating the enhancements separately. Considering that the Raman signal is proportional to the square of dipole moment, ($P = \alpha E$), both enhancement mechanism influences can be viewed as one changes the local electric field (E), and the second changes polarizability (α) near the analyzed molecule. Another way to understand the enhancement of SERS is by looking at the SERS intensity components [20].

$$I_{sers} = I_e N_{sur} \Omega A_e(\omega_e) A_s(\omega_s) \frac{d\sigma}{d\Omega}, \tag{10}$$

where $A_e(\omega_e)$ and $A_s(\omega_s)$ are electromagnetic surface-averaged intensity enhancement factors, I_e excitation light intensity, N_{sur} the number of adsorbed molecules excited by the light, and $\frac{d\sigma}{d\Omega}$ the solid angle of collection optics.

From Eq. (10), it follows that the Raman intensity can be enhanced in three ways:

i. by increasing the number of molecules that are on the metal surface compared to the smooth surface;

ii. by increasing the Raman cross section; and

iii. by increasing the electromagnetic surface averaged intensity enhancement factors.

Experiments have proved that by increasing the surface roughness, the number of absorbed molecules was changed only a few times, leaving us with the last two possibilities. They are electromagnetic (EM) and chemical contributions to the enhancement of Raman signal.

2.1 Chemical enhancement mechanism

CE requires the probe to be chemically bound to the SERS substrate. The CE can be grouped into three contributions to the chemical mechanism: (i) a resonance Raman (RR) effect due to the incident light matching an electronic transition in the molecule (10^3–10^6 contribution), (ii) a charge-transfer (CT) effect where the incident light is in resonance with a metal-molecule or molecule-metal transition. (10–10^4 contribution), (iii) a nonresonant chemical (CHEM) effect due to ground-state orbital overlap between the molecule and the metal (≤10–100 contribution) [21].

RR is a molecular resonance mechanism that arises from the incident light being resonant with a molecule, and without a metal surface, this leads to resonance

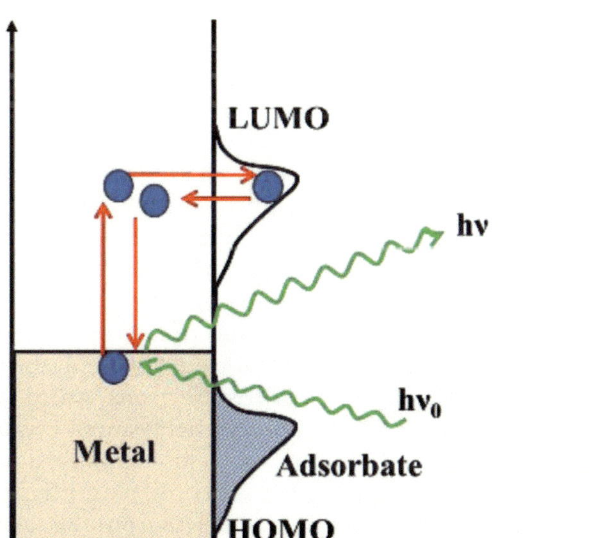

Figure 2.
Schematic diagram of the four-step process of the photon-driven charge transfer model for a molecule adsorbed on an electrode.

Raman scattering. RR involves the formation of a surface complex involving the metal and the analyte, leading to a change in the properties of the molecule (such as the possibility of resonance Raman scattering). The RR effect is typically thought of as a molecular property, and it has been included as a SERS mechanism since the presence of the metal surface can alter where this resonance lies. The CT effect only appears when the molecule and metal are close enough to allow for a sufficient overlap of their wave functions. In this mechanism, tunneling of electrons between the metal and adsorbate molecules takes place. Due to the transfer of an electron from metal to molecule or from molecule to metal, a negative ion is formed. Enhancement occurs when the energy of the negative ion is resonant with the incident photon. This mechanism is explained by considering the molecule and metal system as a whole. It is considered that the Fermi level of the metal layer lies between the molecular ground level and one or more excited states of the molecule. The charge transfer mechanism is short-ranged (0.1–0.5 nm) and strongly dependent on the geometry, bonding, and the molecule energy level [22]. The CHEM effect is the least studied and most difficult to quantify experimentally due to its small contribution to the overall enhancement. The formation of metal-molecular complexes mainly causes the CHEM effect due to chemical bonding [23]. This modifies the ability of the dipole to radiate energy, i.e., it can effectively oppose or amplify the dipole amplitude (**Figure 2**).

2.2 Electromagnetic mechanism

In theory, EM enhancement is analyte independent, while CE is probe-dependent and requires some chemical interaction to the metal surface [24]. Most of SERS enhancements are due to the EM enhancement mechanism and are a direct consequence of the roughness present on the noble metal surfaces [25]. The nanostructure can be formed on the substrate itself or by depositing noble metal nanoparticles. These metal nanoparticles can interact with the excitation light because of unique properties caused by their low dimensions (10–100 nm). The small size of the metal nanoparticles makes a special kind of light-induced electric polarization possible for their surface electrons. Collective oscillations of these electrons, driven by the alternating electric field of the light wave, are called surface plasma oscillations. At a particular frequency, plasmon oscillations are resonant with light; then electric field intensity and Raman scattering from the molecules attached to the nanostructures are enhanced [8, 26]. A locally strong light-induced electric field of plasmons in metal nanoparticle causes the increase of $A_e(\omega_e)$ and $A_s(\omega_s)$ factors. It is because nanoparticles work as a kind of optical antenna, redistributing and concentrating light energy near a nanoparticle. As a result, the cross-section of the light scattering processes, including Raman scattering, can be much larger than the geometrical cross-section of the metal nanoparticle.

The surface electron oscillations in metal nanoparticles can be derived from the classical Drude model, describing metal as a lattice of ions immersed into the “gas” consisting of the free electrons [27, 28]. In a static electric field, the internal field of the metals, generated by the displacement of free electrons, shield the external electric field. As a result, the external electrostatic field cannot create the electric field inside the metal. As a result, if electrostatic fields are applied to the metals, their dielectric permittivity is ambiguous. The dielectric permittivity is the measure of how much the electric field inside a material differs from that of a vacuum. However, when a high-frequency electric field is applied, the free electrons inside the metal cannot completely follow in time with the high-frequency oscillations of the electric field. It creates a situation where at very high frequencies, metal can pass the electric field from the incident light, i.e., behave as a dielectric. The high transparency of these metals in the ultraviolet region can be explained by the fact that they have a lot of free electrons. Electrons of such metals as Al, Cu, Au, and Ag are not completely free but partially bounded.

The surface plasmon frequency ω_{sp} in small spherical metal nanoparticle includes the frequency of the volume plasma ω_p and permittivity of the surrounding dielectric [29]:

$$\omega_{sp} = \frac{\omega_p}{(1+2\varepsilon_d)^{\frac{1}{2}}} \tag{11}$$

Hence at the resonant frequency $\omega_{sp} = \omega$, from Eqs. (11) and (12),

$$\varepsilon_m = -2\varepsilon_d \tag{12}$$

From Eq. (12), it follows that the permittivity of metal should have a negative value. Few metals such as Cu, Ag, and Au exhibit strong visible light plasmon resonance, whereas other transition metals show only broadband in the ultraviolet region. Ag, in particular, is suitable for SERS applications in the visible and near IR because it has a tiny imaginary component in this region and thus is less "lossy" than other metals [30].

When monochromatic radiation of frequency ν_0 and electric filed E interacts with a molecule, it induces a Raman dipole oscillating at a frequency $\mu = \alpha E$. The oscillating Raman dipole radiates a power proportional to $/\mu/^2$ at frequency ν and is the frequency detected as Raman signal in far-field. The same phenomenological description can be applied to SERS. However, the presence of nanostructured metal surface alters the effects in the following ways [31–35]:

a. The electromagnetic field at the metallic surface can be dramatically increased and may result in a possible *local field enhancement*.

b. The radiation properties of the Raman dipole, μ, are affected by the metallic environment and may result in a possible *radiation enhancement*.

3. Applications of Raman spectroscopy

Raman spectrum can give rich information of analyte molecules, and SERS due to its higher signal intensity make it possible to detect analyte molecules in very low concentration, which enhances its practical applications [21]. This technique has a large number of applications in various fields, including trace chemical detection [21, 36], such as dye molecules [37–39], food additives [40, 41], pesticide trace detection [42–44], bioanalysis [45–49], and explosive detection [50, 51]. The detection of a trace amount of hazardous chemicals is also in high demand because of the increasing threat from toxic environments and unreliable food safety [52]. Melamine is a chemical compound and has been widely used in milk and pet food as an additive to increase protein percentage. However, since 2007, melamine, with its contaminant cyanuric acid, has become prominent because of the milk scandal. As a facile and simple spectroscopy technique, SERS has been used to detect melamine content [53, 54]. Apart from this, SERS has been widely used for bioanalysis, i.e., in the detection of biomolecules [55], cancer diagnosis [56, 57], urine component detection [58, 59], and in vivo molecular probing in live cells [60, 61], which play an important role in the life science for health care or treatment. Biomolecules, such as DNA, can also be detected using SERS [62, 63].

3.1 Raman spectroscopy for transition metal dichalcogenides (TMDs)

Transition metal dichalcogenides, as the names suggest, are a class of material that is made up of the transition metals (M = Mo, W, Ta, Pt) and chalcogenides (X = S, Se, Te). The unit cell of bulk MX_2 consists of X–M–X units, where one M plane is sandwiched between two X planes. Depending upon how these units are stacked, different kinds of polytypes are formed, for example:

1. 1T (one X–M–X unit in the unit cell, octahedral coordination, tetragonal symmetry, T stands for trigonal)

2. 2H (two X–M–X layers per repeat unit, trigonal prismatic coordination, hexagonal symmetry, H stands for hexagonal)

3. 3R (three X–M–X layers per repeat unit, trigonal prismatic coordination, rhombohedral symmetry, R stands for rhombohedral)

The schematic diagram of these polytypes is given in **Figure 3**.

In recent years, Raman spectroscopy is adopted to address the challenges in the characterization of these TMDs due to: (1) Many possible structures (generally the

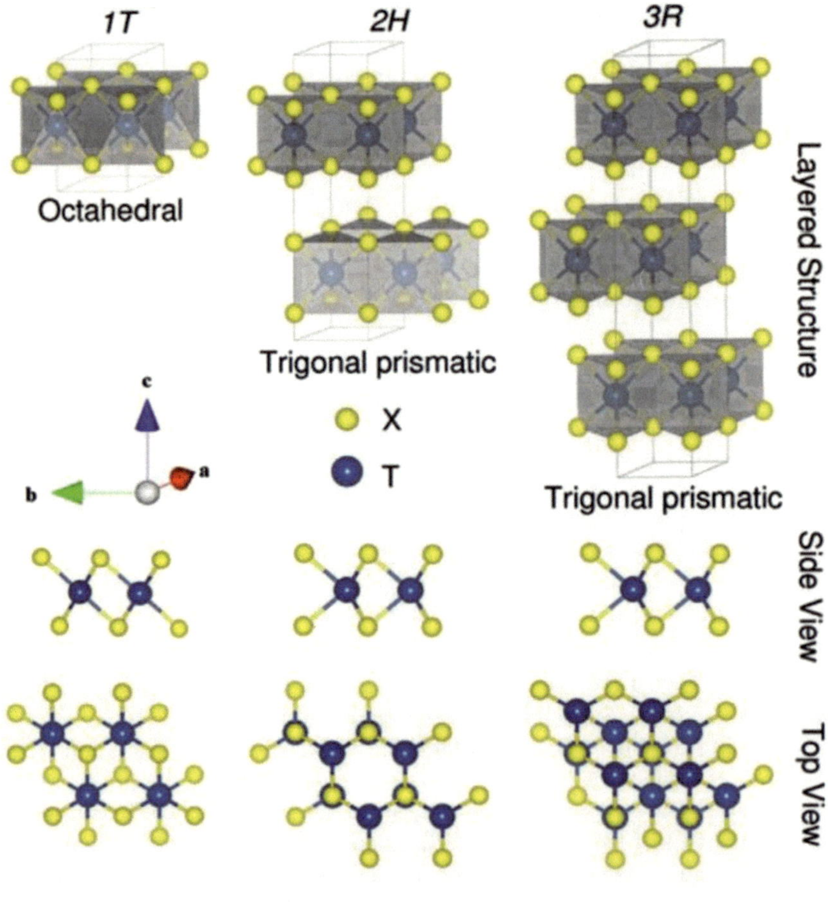

Figure 3.
Structural representation of 1T, 2H, and 3R TMC polytypes and their corresponding metal atom coordination. The side and top view of layered forms are shown [61].

stacking of layers has a rotational (or translational) freedom with reference to an axis perpendicular to (or also along) a 2D plane, giving them non-uniqueness); (2) weak interlayer interaction, which is much smaller than the intralayer interactions; and (3) observation of inner layers (the physical properties of each layer within the stacked TMDs are not selectively accessible except for the outermost layer). Other than resolving these challenges, Raman spectroscopy can identify functional groups, structural damage, unwanted by-products, and chemical modifications introduced during synthesis, processing, or its placement on the various substrates during device fabrication. It is spectroscopy, which is nondestructive, quick, and noninvasive for characterizing the TMD materials with high selectivity [64–66].

In general, the Raman spectroscopy has been widely used to determine the layers of the TMDs. For all the layered materials, which also includes TMDs, there are typically two categories of Raman vibrations. One is intralayer vibrations, which occur within a layer and normally appear in the high-frequency region of the spectra. The second category of Raman vibrations is observed due to the relative motion of the layers. These vibrations give the interlayer Raman modes, which are normally observed in the low-frequency region of the spectra ($<100\ cm^{-1}$).

In the layered TMDs, to determine the number of layers N for few-layer TMDs, high-frequency intralayer Raman modes can be used [67, 68]. For example, the layer number (N) dependence of the peak position and width for E^1_{2g} and A_{1g} of NL-MoS_2 is shown in **Figure 4(a)**. The A1g and E^1_{2g} modes undergo blue and redshifts with increasing numbers of layers for MoS_2.

The position of (E^1_{2g}) and (A_{1g}) with decreasing thickness from the bulk to 1 L shows opposite trends, as shown in **Figure 4(b)**.

However, these two modes would decrease in frequency from 2L to 1L based on the linear chain model (only van der Waals interactions are included). This unexpected behavior of E^1_{2g} suggests that interactions other than van der Waals forces also exist [68–70], which is also disclosed by the anomalous Davydov splitting between E^1_{2g} and E_{1u} [71].

Molina-Sanchez et al. [72] carefully examined and reported the relationship between the number of monolayers and the Raman active modes (A_{1g} and E^1_{2g}). They demonstrated that the weak interlayer interaction is the leading cause of the frequency increase (i.e., for A_{1g}) with the number of layers. Moreover, the decrease

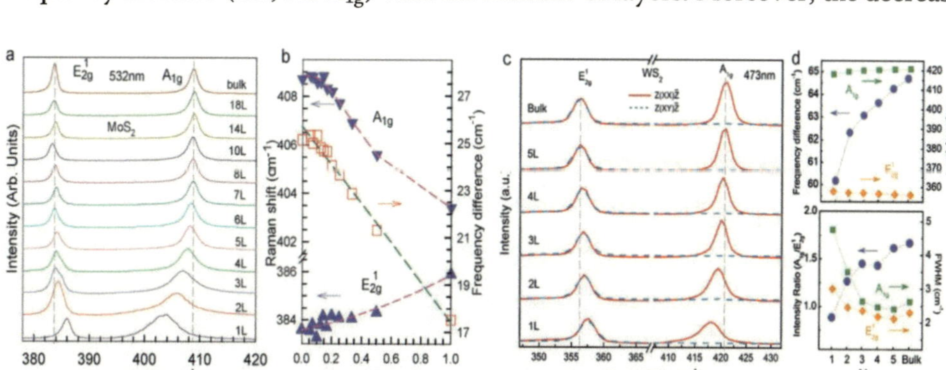

Figure 4.
(a) Raman spectra of NL-(N = 1–8, 10, 14 and 18) and bulk MoS_2. The two gray-dashed lines indicate Pos (E^1_{2g} E1 2g) and Pos(A_{1g}) in bulk MoS_2. (b) Frequency (ω) of E^1_{2g} and A_{1g} and the frequency difference (Δω) between E^1_{2g} and A_{1g} as a function of 1/N. For $1 \leq N \leq 5$, the linear fitting gives $\Delta\omega(A–E) = 25.8–8.4/N$. (c) Polarized Raman spectra of 1–5L and bulk WS_2, with the frequencies of E^1_{2g} and A_{1g} frequency difference, I (A_{1g})/I(E^1_{2g}), and peak width summarized in (d) [66, 73].

in the E_{2g}^1 phonon frequency is associated with a stronger dielectric screening of the long-range Coulomb interaction (which is induced by the effective charges resulting from the relative displacement between Mo and S atoms) in a few layers and in bulk. Thus, it is expected to exhibit an anomalous frequency trend in which the A_{1g} mode increases in frequency with an increasing number of layers, while the E_{2g}^1 mode decreases.

In the case of the WS_2, the proximity of the 2LA(M) and E_{2g}^1 peaks makes it challenging to measure the accurate difference between the A_{1g} and E_{2g}^1 peaks and increases the chance of error in determining the frequency shift of both the modes. To resolve this problem, Berkdemir et al. [74] reported a new method based on the ratio of the intensities of I_{2LA} and I_{A1g} peaks. In this method, the authors reported that the absolute intensity of the 2LA(M) mode increases with decreasing the number of layers, while the intensity of the A_{1g} displays the opposite behavior [74]. The behavior of the A_{1g} mode with a decreasing number of layers presumably results from weaker interlayer contributions to the phonon restoring forces.

3.2 Application of Raman spectroscopy in photovoltaics

Conducting polymers are widely used in organic light-emitting diodes, heterojunction diodes, organic thin-film transistors, solar cells, actuators, sensors, etc. [75–79]. Poly(3,4-ethylenedioxythiophene)-poly(styrene sulfonate) (PEDOT: PSS) is extensively used conducting polymer because of its high conductivity, excellent thermal stability, transparency, structural stability. PEDOT:PSS polymer is a promising candidate as a transparent electrode for optoelectronic devices. The solvent treatment of PEDOT:PSS films may affect the conformation of the polymer. The structure of the PEDOT chain changes from benzoid to quinoid structure after solvent treatment [75, 80]. The effect of the conformation of the PEDOT chains in the PEDOT:PSS film before and after the dimethyl sulfoxide (DMSO) treatment was studied by Raman spectroscopy. **Figure 5** shows the Raman spectra of PEDOT: PSS with different concentrations of DMSO. The most obvious change was observed for the strongest band between 1400 and 1500 cm^{-1}. The highly

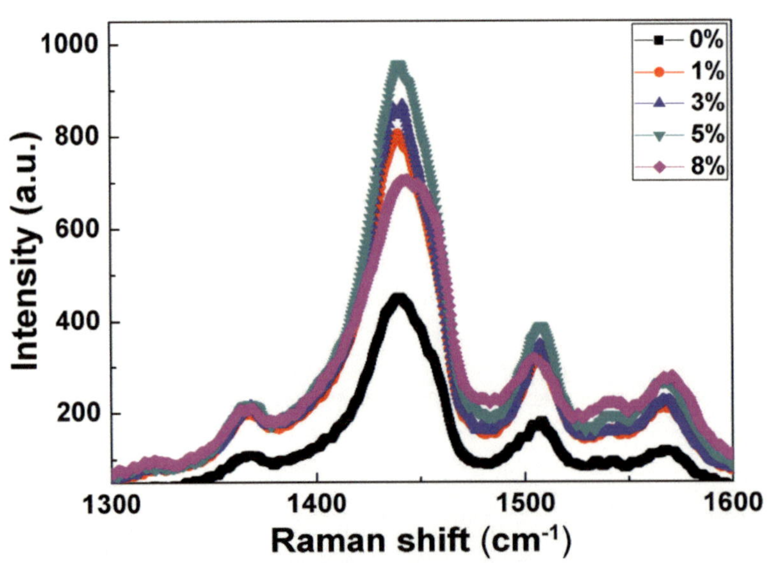

Figure 5.
Raman spectra of PEDOT:PSS films with 0–8 vol.% DMSO.

conductive PEDOT:PSS film exhibits a narrower band. This change is similar to that of PEDOT:PSS film treated with ethylene glycols reported by Xia et al. [81]. These vibrational modes correspond to the stretching vibrations of $C_\alpha = C_\beta$ on the five-member ring of PEDOT.

The band at 1440 cm^{-1} is associated with the $C_\alpha{=}C_\beta$ symmetric vibration. The band near about 1368 cm^{-1} is associated with the $C_\beta{-}C_\beta$ stretching. Raman peaks located at 1508 and 1568 cm^{-1} are associated with the $C_\alpha{=}C_\beta$ asymmetric stretching vibrations. The band at 1540 cm^{-1} has been related to the splitting of these asymmetrical stretching vibrations [79, 81–84]. Two kinds of resonant structures have been proposed for PEDOT, namely, benzoid and quinoid structure. For coil conformation, benzoid structure is the favorite structure, and quinoid structure is the favorite structure for linear and expanded-coil structure. Both benzoid and quinoid resonant structures exist simultaneously in pristine PEDOT:PSS film. The benzoid structure may be transformed into the quinoid structure after DMSO treatment so that quinoid structure becomes dominant in the highly conductive PEDOT:PSS film. The conducting PEDOT:PSS films are vastly used in optoelectronic devices.

4. SERS substrates

"SERS substrates" are any nanostructured metallic platform that supports plasmon resonance and amplifies Raman signals [85]. Herein, SERS substrates are classified into two broad distinctions:

a. **Random morphology SERS substrates** include roughened electrodes, metallic silver and gold colloids, metal-island film on planer substrate, and other related substrates.

b. **Ordered or periodic metallic SERS substrates** include arrays of regular morphology metallic nanotextures created on planar substrates using nanolithography and other physical vapor deposition techniques.

4.1 Random morphology SERS substrates

Random morphology SERS substrates are inhomogeneous and are not highly reproducible [86]. Roughened electrodes are the most primitive SERS substrate and were discovered by Fleischmann et al. [9]. These substrates are typically created by running the redox cycle in an electrochemical cell containing a metallic salt solution. Such substrates have gained popularity due to an ability to adjust electrode potential to understand the charge transfer phenomenon between adsorbate and metallic surface [87]. Regardless, the importance of this substrate is decreasing substantially due to relatively low enhancement factors.

Among the random morphology SERS substrates, silver or gold colloids are the most common substrates used in both early and more recent studies. Since colloids are easy to produce in a laboratory and tend to generate large enhancement factors, most researchers are still involved in colloid-based SERS rather than more sophisticated substrates [88, 89]. Metallic colloids are also of historic significance related to SERS development, as the first single-molecule SERS detection was reported using colloid substrates [90]. In colloid-based methods, nanoparticle size and geometry can be controlled by altering experimental conditions. One of the most popular methods for controlling nanoparticle morphology stems from the polyol synthesis of silver nanocubes by Sun and Xia [78]. In addition to the nanocubes, various groups have produced octahedra and cuboctahedra [91] and octapods [92].

El-Sayed et al. have contributed to the control of particle morphology, yielding a variety of interesting and useful structures [93–95]. Since the size, shape, and material of the particles govern the resulting plasmonic resonance characteristics, significant effort has been exerted in the control of plasmon resonance via core-shell and alloyed particles, to which the Halas group has a large contribution [96–98].

SERS substrates have also been fabricated by depositing nanoparticles onto different surfaces. These simple deposition approaches include micro pipetting [99], soaking [100], screen-printing [101], filtration [102], and inkjet printing [103]. However, a major concern with these simple deposition processes is that the hotspots are generated and distributed randomly over the substrate. The differences in metallic particle sizes and their shapes due to the differences in preparation recipes can lead to several orders of magnitude difference in the SERS enhancement factor.

In order to obtain more consistent hotspots from nanoparticles, researchers have explored both self-assembly [99] and directed assembly techniques, such as the Langmuir-Blodgett techniques [91], to create regular arrays of nanoparticles. However, these techniques also introduce more complexity to the fabrication process.

4.2 Periodic or uniform SERS substrate

Although metallic colloidal particles are known for their high SERS EF and possibility to accomplish SERS spectra of a single molecule, it is often challenging to reproduce or routinely deliver such a high performing SERS feature. To overcome this issue, a few alternatives have been introduced on engineering periodic arrays of metallic nanostructured SERS substrates. Nanosphere lithography (NSL), developed by Van Duyne et al., is one of the most extensively used nano-fabrication procedures used in understanding SERS phenomena and performing plasmonic-based sensing [104–106]. It involves the assembly of polystyrene nanospheres into a regular array, using this as a mask to create periodic nanostructures, sometimes called "metal island films," by the evaporation of Ag or Au through the gaps created by the packing of the nanospheres, followed by removal of the nanospheres. A variation on this method, called "metal film over nanospheres," is to evaporate a metal film directly onto the nanosphere template, using closely-packed nanospheres to pattern the substrate surface itself [107]. Electron beam lithography (EBL) is another most widely used conventional nanofabrication technique in designing uniform and controlled morphology SERS substrates [108, 109]. In the literature, there are many SERS-active surface designs prepared with EBL [110, 111]. These designs are mainly periodic arrays of simple nano-structures, and generally, the relation between the LSP resonance wavelength and SERS signal enhancement is studied. For example, Le Ru et al. have taken SERS measurements, from periodic gold dot, square, and triangle arrays, from "Rhodamine 6G" [112]. They demonstrated that the localized plasmon resonances, which are at the origin of visible-NIR extinction spectra and the SERS effect, can be tuned to any desired wavelength by varying the particle shape/size and spacing, thus tuning the Raman amplification. In a very similar study, Gunnarsson et al. studied similar Ag structures on a silicon wafer for the same molecule and reported that better results are obtained than nano-roughened Ag film [110]. They investigated the size and geometry dependence of the SERS effect on supported particles, by manufacturing artificial structures by modern nanofabrication techniques. Arrays of 100–200 nm silver particles of different shapes were prepared on a Si wafer by electron beam lithography. Kahl et al. have shown that the SERS measurements of "Rhodamine 6G" on gold periodic nano-dot arrays and grating structures resulted in order of magnitude better SERS enhancement when compared with metal-island film substrates [113]. They demonstrated two different methods of substrate fabrication by e-beam lithography. In the first method, regular fields of nanoparticles are produced by the lift-off technique. A silver layer is evaporated on the structured resist, and the resist is removed afterward. In the second method, gratings or crossed gratings are transferred into a silicon wafer with a thermal oxide surface layer by reactive ion etching (RIE). Then, the e-beam resist is removed, and finally, a silver layer is evaporated. Hatab et al. have demonstrated significant SERS enhancement factors exceeding 10^{11}, resulting from a new configuration of elevated gold bowtie nano-antenna arrays with optimized array periodicity [114]. A process combining nanofabrication steps of pattern definition by EBL, metal deposition, lift-off, and RIE arranged in a particular sequence was used to fabricate the elevated gold bowtie arrays on Si wafers. The elevated bowties allow the manifestation of intrinsic plasmonic coupling effects in suspended nanocavities, or the tip-to-tip nanogaps, from structures that are not in physical contact with a substrate. This configuration results in up to two orders of additional magnitude enhancement in SERS response compared to that of nonelevated bowtie arrays. The diversity of designs is endless when fabrication with EBL is considered. However, these techniques, while excellent at making SERS substrates with defined characteristics, are hampered by the slow, serial nature and high cost of the processes used in their fabrication.

4.3 SERS from silver columnar film

The SERS enhancement strongly depends on the substrate. As already discussed earlier, various techniques have been proposed and identified for the fabrication of the SERS substrate. However, only a few methods are available to develop uniform, reproducible, robust, stable, and cost-effective SERS substrates. Recently, silver columnar thin films fabricated by glancing angle deposition (GLAD) have been identified as high sensitivity SERS active substrates [115–122]. A remarkable SERS enhancement factor with applications in sensing the biomolecules at very low concentrations has been observed on the silver nanorod arrays [123]. To understand the SERS mechanism and attain a maximum possible enhancement, large numbers of studies have been performed on the Ag nanorod (AgNR) arrays. In an interesting study, Chaney et al. have investigated the SERS response as a function of the nanorod length using trans-1,2-bis(4-pyridyl)ethane (BPE) as a probe molecule at an excitation wavelength of 785 nm [124]. They found that the SERS intensity increases dramatically with nanorod length. Zhou et al. fabricated aligned, single-crystalline AgNRs on planar Si substrates by GLAD technique, with sample substrate cooled by liquid nitrogen in the e-beam deposition system [125]. They were successful in detecting aqueous solution of10^{-12} $molL^{-1}$ Rhodamine 6G by the porous Ag film with nanorods. They also deposited AgNRs on Ag, Al, Si, and Ti thin films with a thickness of 100 and 400 nm, respectively, to achieve layers with different reflectivities. The SERS intensity of the AgNRs grown on Ag thin film was found to be higher than others, and the SERS intensity of the AgNRs on Al film was larger than that on Ti film, and the AgNRs on Si film showed the minimum SERS intensity. They concluded that the larger the under-layer reflectivity, the larger the SERS performance of substrate. So, the pre-deposition of Ag layer under AgNRs can be an effective way to promote the SERS performance of AgNRs. Zhang et al. have made AgNRs in film grow into periodic patterns at a micro-nano scale, and they showed that the AgNR film with periodic patterns exhibits better SERS performance than Ag film with nanorods arranged randomly as before [126]. He et al. also reported a new scalable strategy based on dynamic shadowing growth (DSG) to fabricate large-scale chiral Swiss roll nanostructures. They developed a chiral conical Swiss roll nanostructure by helically stacking Ag films on a SiO_2 frustum with SiO_2 films as insulating layers [127]. They also showed that the chiral dichroism (CD) spectral feature can be tuned by changing the bead diameter. They achieved a broadband CD response in visible to near-IR region by making the bead diameter a few hundred nanometers. Mark et al. combined the low-temperature shadow deposition with nanoscale patterning to fabricate nanocolloids with anisotropic three-dimensional shapes, feature sizes down to 20 nm [128]. They first deposited a uniform hexagonal array of Au nanodots deposited onto a Si wafer by micellar nanolithography. Then they deposited material onto the substrate by physical vapor deposition at grazing incidence. To reduce the adatoms' mobility and reduce the diffuse during growth, they cooled the substrate. So, by combining the uniform nano-seeding and low-temperature growth, they fabricated various complex hybrid nanostructures of many materials like from Al_2O_3, Ti, and Cu. GLAD has also emerged as a powerful tool for the fabrication of 3D chiral plasmonic nanostructures. Titus et al. investigated the optical properties of Ti-doped Ag helices in the visible and near-infrared ranges using transmission ellipsometry and spectroscopy fabricated by GLAD [129]. Nair et al. reported the fabrication of wafer-scale 3D chiral nanoplasmonic substrates with different dielectric templates, namely, silica, magnesium fluoride, and titanium dioxide using GLAD [130]. They have also investigated the effect of interparticle separation on the chiroptical response of chiral nanohelices [131].

Hence, we can see that the development of fabrication and application of substrates for SERS is driven by nanotechnology and the development of high-end fabrication processes. Increasingly SERS substrates with high sensitivity and reproducibility are invented by electrochemical deposition, physical vapor deposition of the metal film, metal nanoparticle colloids, and so forth and applied into various fields, such as detection of pollutants at trace level, surface analysis, biomolecule, and bacteria detection. With the development of SERS substrate, advancement in Raman spectrometers, and tip-enhanced Raman scattering (as the combination of SERS and atomic force microscopy), SERS is becoming increasingly popular as a detection and diagnostic tool.

Author details

Samir Kumar[1*], Prabhat Kumar[2], Anamika Das[3] and Chandra Shakher Pathak[4]

1 Department of Micro Engineering, Graduate School of Engineering, Kyoto University, Kyoto, Japan

2 Department of Thin Films and Nanostructures, Institute of Physics of the Czech Academy of Sciences, Prague, Czech Republic

3 Department of Paramedical Sciences, Guru Kashi University, Bhatinda, India

4 Department of Solar Energy and Environmental Physics, Ben-Gurion National Solar Energy Center, Jacob Blaustein Institutes for Desert Research, Ben-Gurion University of the Negev, Midreshet Ben-Gurion, Israel

*Address all correspondence to: drsamirkumar2017@gmail.coom

IntechOpen

© 2020 The Author(s). Licensee IntechOpen. This chapter is distributed under the terms of the Creative Commons Attribution License (http://creativecommons.org/licenses/by/3.0), which permits unrestricted use, distribution, and reproduction in any medium, provided the original work is properly cited.

References

[1] Weber WH, Merlin R. Raman Scattering in Materials Science. Berlin, Heidelberg: Springer; 2000

[2] Smekal A. Zur Quantentheorie der Dispersion. Die Naturwissenschaften. 1923;**11**:873-875. DOI: 10.1007/BF01576902

[3] Krishnan KS, Raman CV. A new class of spectra due to secundary radiation. Indian Journal of Physics. 1928;**2**: 399-419

[4] Krishnan RS, Shankar RK. Raman effect: History of the discovery. Journal of Raman Specroscopy. 1981;**10**:1-8. DOI: 10.1002/jrs.1250100103

[5] Schrader B. Infrared and Raman Spectroscopy: Methods and Applications. Weinheim (Federal Republic of Germany): VCH; 1995

[6] Long DA. Quantum mechanical theory of Rayleigh and Raman scattering. In: Long DA, editor. The Ram Effect. Chichester, UK: John Wiley & Sons, Ltd.; 2002. pp. 49-84. DOI: 10.1002/0470845767.ch4

[7] Smith E, Dent G. Introduction, basic theory and principles. In: Smith E, Dent G, editors. Modern Raman Spectrosopy—A Practical Approach. Vol. 5. 2005. pp. 1-21. DOI: 10.1002/0470011831.ch1

[8] Moskovits M. Surface-enhanced spectroscopy. Reviews of Modern Physics. 1985;**57**:783-826. DOI: 10.1103/RevModPhys.57.783

[9] Fleischmann M, Hendra PJ, McQuillan AJ. Raman spectra of pyridine adsorbed at a silver electrode. Chemical Physics Letters. 1974;**26**: 163-166. DOI: 10.1016/0009-2614(74)85388-1

[10] Jeanmaire DL, Van Duyne RP. Surface Raman spectroelectrochemistry. Part I. Heterocyclic, aromatic, and aliphatic amines adsorbed on the anodized silver electrode. Journal of Electroanalytical Chemistry. 1977;**84**: 1-20. DOI: 10.1016/S0022-0728(77)80224-6

[11] Albrecht MG, Creighton JA. Intense Raman spectra at a roughened silver electrode. Electrochimica Acta. 1978;**23**: 1103-1105. DOI: 10.1016/0013-4686(78)85022-1

[12] Le Ru EC, Etchegoin PG. Quantifying SERS enhancements. MRS Bulletin. 2013;**38**:631-640. DOI: 10.1557/mrs.2013.158

[13] Bochenkov V, Baumberg J, Noginov M, Benz F, Aldewachi H, Schmid S, et al. Applications of plasmonics: General discussion. Faraday Discussions. 2015;**178**:435-466. DOI: 10.1039/C5FD90025E

[14] Kneipp K, Moskovits M, Kneipp H. Surface-Enhanced Raman Scattering: Physics and Applications. Berlin, Heidelberg: Springer; 2006

[15] Le Ru EC, Etchegoin PG, Meyer M. Enhancement factor distribution around a single surface-enhanced Raman scattering hot spot and its relation to single molecule detection. The Journal of Chemical Physics. 2006;**125**:204701. DOI: 10.1063/1.2390694

[16] Nardou E, Vouagner D, Jurdyc AM, Berthelot A, Pillonnet A, Sablonire V, et al. Distance dependence of the surface enhanced Raman scattering effect observed in amorphous TiO_2 on nanostructured gold. Optical Materials. 2011;**33**:1907-1910. DOI: 10.1016/j.optmat.2011.03.024

[17] Zhao Y, Liu X, Lei DY, Chai Y. Effects of surface roughness of Ag thin films on surface-enhanced Raman

spectroscopy of graphene: Spatial nonlocality and physisorption strain. Nanoscale. 2014;**6**:1311-1317. DOI: 10.1039/c3nr05303b

[18] McNay G, Eustace D, Smith WE, Faulds K, Graham D. Surface-enhanced Raman scattering (SERS) and surface-enhanced resonance Raman scattering (SERRS): A review of applications. Applied Spectroscopy. 2011;**65**:825-837. DOI: 10.1366/11-06365

[19] Su J-P, Lee Y-T, Lu S-Y, Lin JS. Chemical mechanism of surface-enhanced Raman scattering spectrum of pyridine adsorbed on Ag cluster: *Ab initio* molecular dynamics approach. Journal of Computational Chemistry. 2013;**34**:2806-2815. DOI: 10.1002/jcc.23464

[20] Kruszewski S. Enhancement mechanisms in the SERS phenomenon. In: Proceedings SPIE 3320, Tenth Polish-Czech-Slovak Optical Conference: Wave and Quantum Aspects of Contemporary Optics. 1 January 1998. DOI: 10.1117/12.301353

[21] Schlücker S. Surface-enhanced Raman spectroscopy: Concepts and chemical applications. Angewandte Chemie, International Edition. 2014;**53**: 4756-4795. DOI: 10.1002/anie.201205748

[22] Ueba H, Ichimura S, Yamada H. Where are we in the study of SERS? Role of chemisorption and charge transfer. Surface Science. 1982;**119**:433-448. DOI: 10.1016/0039-6028(82)90309-0

[23] Saikin SK, Olivares-Amaya R, Rappoport D, Stopa M, Aspuru-Guzik A. On the chemical bonding effects in the Raman response: Benzenethiol adsorbed on silver clusters. Physical Chemistry Chemical Physics. 2009;**11**: 9401-9411. DOI: 10.1039/b906885f

[24] Otto A, Mrozek I, Grabhorn H, Akemann W. Surface-enhanced Raman scattering. Journal of Physics. Condensed Matter. 1992;**4**:1143-1212. DOI: 10.1088/0953-8984/4/5/001

[25] Saikin SK, Chu Y, Rappoport D, Crozier KB, Aspuru-Guzik A. Separation of electromagnetic and chemical contributions to surface-enhanced Raman spectra on nanoengineered plasmonic substrates. Journal of Physical Chemistry Letters. 2010;**1**:2740-2746. DOI: 10.1021/jz1008714

[26] Kneipp K. Surface-enhanced Raman scattering. Physics Today. 2007;**60**: 40-46. DOI: 10.1063/1.2812122

[27] Dressel M, Scheffler M. Verifying the Drude response. Annalen der Physik. 2006;**15**:535-544. DOI: 10.1002/andp.200510198

[28] Mahmudin L, Suharyadi E, Bambang A, Utomo S, Abraha K. Optical properties of silver nanoparticles for surface plasmon resonance (SPR)-based biosensor applications. Journal of Modern Physics. 2015;**6**:1071-1076. DOI: 10.4236/jmp.2015.68111

[29] Gordon R. Surface plasmon nanophotonics: A tutorial. IEEE Nanotechnology Magazine. 2008;**2**: 12-18. DOI: 10.1109/MNANO.2008.931481

[30] Alsawafta M, Wahbeh M, Van Truong V. Plasmonic modes and optical properties of gold and silver ellipsoidal nanoparticles by the discrete dipole approximation. Journal of Nanomaterials. 2012;**2012**. DOI: 10.1155/2012/457968

[31] Deng CY, Zhang GL, Zou B, Shi HL, Liang YJ, Li YC, et al. Local electric field enhancement of neighboring Ag nanoparticles in surface enhanced Raman scattering. Advances in Materials Research. 2013;**760–762**: 801-805. DOI: 10.4028/www.scientific.net/AMR.760-762.801

[32] Hexter RM, Albrecht MG. Metal surface Raman spectroscopy: Theory. Spectrochimica Acta Part A: Molecular Spectroscopy. 1979;**35**:233-251. DOI: 10.1016/0584-8539(79)80143-9

[33] Cialla D, Petschulat J, Hübner U, Schneidewind H, Zeisberger M, Mattheis R, et al. Investigation on the second part of the electromagnetic SERS enhancement and resulting fabrication strategies of anisotropic plasmonic arrays. ChemPhysChem. 2010;**11**: 1918-1924. DOI: 10.1002/cphc.200901009

[34] Gersten JI, Nitzan A. Electromagnetic theory: A spheroidal model. In: Chang RK, Furtak TE, editors. Surface Enhanced Raman Scattering. Boston, MA, US: Springer; 1982. pp. 89-107. DOI: 10.1007/978-1-4615-9257-0_5

[35] Zeman EJ, Schatz GC. An accurate electromagnetic theory study of surface enhancement factors for silver, gold, copper, lithium, sodium, aluminum, gallium, indium, zinc, and cadmium. The Journal of Physical Chemistry. 1987; **91**:634-643. DOI: 10.1021/j100287a028

[36] Zhang L, Xu J, Mi L, Gong H, Jiang S, Yu Q. Multifunctional magnetic-plasmonic nanoparticles for fast concentration and sensitive detection of bacteria using SERS. Biosensors & Bioelectronics. 2012;**31**: 130-136. DOI: 10.1016/j.bios.2011.10.006

[37] Zhao X, Zhang B, Ai K, Zhang G, Cao L, Liu X, et al. Monitoring catalytic degradation of dye molecules on silver-coated ZnO nanowire arrays by surface-enhanced Raman spectroscopy. Journal of Materials Chemistry. 2009;**19**: 5547-5553. DOI: 10.1039/B902883H

[38] Prikhodko SV, Rambaldi DC, King A, Burr E, Muros V, Kakoulli I. New advancements in SERS dye detection using interfaced SEM and Raman spectromicroscopy (μRS). Journal of Raman Specroscopy. 2015;**46**: 632-635. DOI: 10.1002/jrs.4710

[39] Kumar S, Lodhi DK, Singh JP. Highly sensitive multifunctional recyclable Ag-TiO_2 nanorod SERS substrates for photocatalytic degradation and detection of dye molecules. RSC Advances. 2016;**6**: 45120-45126. DOI: 10.1039/C6RA06163J

[40] Granger JH, Schlotter NE, Crawford AC, Porter MD. Prospects for point-of-care pathogen diagnostics using surface-enhanced Raman scattering (SERS). Chemical Society Reviews. 2016;**45**:3865-3882. DOI: 10.1039/C5CS00828J

[41] Craig AP, Franca AS, Irudayaraj J. Surface-enhanced Raman spectroscopy applied to food safety. Annual Review of Food Science and Technology. 2013;**4**: 369-380. DOI: 10.1146/annurev-food-022811-101227

[42] Wang P, Wu L, Lu Z, Li Q, Yin W, Ding F, et al. Gecko-inspired nanotentacle SERS substrate for rapid sampling and reliable detection of pesticide residues in fruits and vegetables. Analytical Chemistry. 2017; **89**:2424-2431. DOI: 10.1021/acs.analchem.6b04324

[43] Furini LN, Constantino CJL, Sanchez-Cortes S, Otero JC, López-Tocón I. Adsorption of carbendazim pesticide on plasmonic nanoparticles studied by surface-enhanced Raman scattering. Journal of Colloid and Interface Science. 2016;**465**:183-189. DOI: 10.1016/j.jcis.2015.11.045

[44] Kumar S, Goel P, Singh JP. Flexible and robust SERS active substrates for conformal rapid detection of pesticide residues from fruits. Sensors and Actuators B: Chemical. 2017;**241**: 577-583. DOI: 10.1016/j.snb.2016.10.106

[45] Jarvis RM, Goodacre R. Discrimination of bacteria using surface-enhanced Raman spectroscopy. Analytical Chemistry. 2004;**76**:40-47. DOI: 10.1021/ac034689c

[46] Wu X, Huang YW, Park B, Tripp RA, Zhao Y. Differentiation and classification of bacteria using vancomycin functionalized silver nanorods array based surface-enhanced Raman spectroscopy and chemometric analysis. Talanta. 2015; **139**:96-103. DOI: 10.1016/j.talanta.2015.02.045

[47] Xu K, Zhou R, Takei K, Hong M. Toward flexible surface-enhanced Raman scattering (SERS) sensors for point-of-care diagnostics. Advancement of Science. 2019;**6**:1900925. DOI: 10.1002/advs.201900925

[48] Kumar S, Lodhi DK, Goel P, Neeti N, Mishra P, Singh JP. A facile method for fabrication of buckled PDMS silver nanorod arrays as active 3D SERS cages for bacterial sensing. Chemical Communications. 2015;**51**: 12411-12414. DOI: 10.1039/C5CC03604F

[49] Kumar S, Namura K, Suzuki M. Proposal for a gel-based SERS sensor. Proceedings of SPIE. 2019;**10894**: 1089414. DOI: 10.1117/12.2506951

[50] Hakonen A, Andersson PO, Stenbæk Schmidt M, Rindzevicius T, Käll M. Explosive and chemical threat detection by surface-enhanced Raman scattering: A review. Analytica Chimica Acta. 2015;**893**:1-13. DOI: 10.1016/j.aca.2015.04.010

[51] Fierro-Mercado PM, Hernández-Rivera SP. Highly sensitive filter paper substrate for SERS trace explosives detection. International Journal of Spectroscopy. 2012;**716527**:7. Available from: http://www.hindawi.com/journals/ijs/2012/716527/

[52] Hakonen A, Wang F, Andersson PO, Wingfors H, Rindzevicius T, Schmidt MS, et al. Hand-held femtogram detection of hazardous picric acid with hydrophobic Ag nanopillar SERS substrates and mechanism of elasto-capillarity. ACS Sensors. 2017;**2**:198-202. DOI: 10.1021/acssensors.6b00749

[53] Giovannozzi AM, Rolle F, Sega M, Abete MC, Marchis D, Rossi AM. Rapid and sensitive detection of melamine in milk with gold nanoparticles by surface enhanced Raman scattering. Food Chemistry. 2014;**159**:250-256. DOI: 10.1016/j.foodchem.2014.03.013

[54] Du X, Chu H, Huang Y, Zhao Y. Qualitative and quantitative determination of melamine by surface-enhanced Raman spectroscopy using silver nanorod array substrates. Applied Spectroscopy. 2010;**64**:781-785

[55] Hughes J, Izake EL, Lott WB, Ayoko GA, Sillence M. Ultra sensitive label free surface enhanced Raman spectroscopy method for the detection of biomolecules. Talanta. 2014;**130**: 20-25. DOI: 10.1016/j.talanta.2014.06.012

[56] Kong KV, Leong WK, Lam Z, Gong T, Goh D, Lau WKO, et al. A rapid and label-free SERS detection method for biomarkers in clinical biofluids. Small. 2014;**10**:5030-5034. DOI: 10.1002/smll.201401713

[57] Zhu J, Zhou J, Guo J, Cai W, Liu B, Wang Z, et al. Surface-enhanced Raman spectroscopy investigation on human breast cancer cells. Chemistry Central Journal. 2013;7:37. DOI: 10.1186/1752-153X-7-37

[58] Dong R, Weng S, Yang L, Liu J. Detection and direct readout of drugs in human urine using dynamic surface-enhanced Raman spectroscopy and support vector machines. Analytical

Chemistry. 2015;**87**:2937-2944. DOI: 10.1021/acs.analchem.5b00137

[59] Villa JEL, Poppi RJ. A portable SERS method for the determination of uric acid using a paper-based substrate and multivariate curve resolution. The Analyst. 2016;**141**:1966-1972. DOI: 10.1039/c5an02398j

[60] Willets KA. Surface-enhanced Raman scattering (SERS) for probing internal cellular structure and dynamics. Analytical and Bioanalytical Chemistry. 2009;**394**:85-94. DOI: 10.1007/s00216-009-2682-3

[61] Radziuk D, Moehwald H. Prospects for plasmonic hot spots in single molecule SERS towards the chemical imaging of live cells. Physical Chemistry Chemical Physics. 2015;**17**:21072-21093. DOI: 10.1039/C4CP04946B

[62] Lim D-K, Jeon K-S, Hwang J-H, Kim H, Kwon S, Suh YD, et al. Highly uniform and reproducible surface-enhanced Raman scattering from DNA-tailorable nanoparticles with 1-nm interior gap. Nature Nanotechnology. 2011;**6**:452-460. DOI: 10.1038/nnano.2011.79

[63] Harper MM, McKeating KS, Faulds K. Recent developments and future directions in SERS for bioanalysis. Physical Chemistry Chemical Physics. 2013;**15**:5312-5328. DOI: 10.1039/c2cp43859c

[64] Zhang X, Tan QH, Bin Wu J, Shi W, Tan PH. Review on the Raman spectroscopy of different types of layered materials. Nanoscale. 2016;**8**: 6435-6450. DOI: 10.1039/c5nr07205k

[65] Saito R, Tatsumi Y, Huang S, Ling X, Dresselhaus MS. Raman spectroscopy of transition metal dichalcogenides. Journal of Physics. Condensed Matter. 2016;**28**:353002. DOI: 10.1088/0953-8984/28/35/353002

[66] Zhang X, Qiao X-F, Shi W, Wu J-B, Jiang D-S, Tan P-H. Phonon and Raman scattering of two-dimensional transition metal dichalcogenides from monolayer, multilayer to bulk material. Chemical Society Reviews. 2015;**44**:2757-2785. DOI: 10.1039/C4CS00282B

[67] Li H, Zhang Q, Yap CCR, Tay BK, Edwin THT, Olivier A, et al. From bulk to monolayer MoS_2: Evolution of Raman scattering. Advanced Functional Materials. 2012;**22**:1385-1390. DOI: 10.1002/adfm.201102111

[68] Lee C, Yan H, Brus LE, Heinz TF, Hone J, Ryu S. Anomalous lattice vibrations of single- and few-layer MoS_2. ACS Nano. 2010;**4**:2695-2700. DOI: 10.1021/nn1003937

[69] Molina-Sánchez A, Wirtz L. Phonons in single-layer and few-layer MoS_2 and WS_2. Physical Review B. 2011; **84**:155413. DOI: 10.1103/PhysRevB.84.155413

[70] Luo X, Zhao Y, Zhang J, Xiong Q, Quek SY. Anomalous frequency trends in MoS_2 thin films attributed to surface effects. Physical Review B. 2013;**88**: 075320. DOI: 10.1103/PhysRevB.88.075320

[71] Verble JL, Wieting TJ. Lattice mode degeneracy in MoS_2 and other layer compounds. Physical Review Letters. 1970;**25**:362-365. DOI: 10.1103/PhysRevLett.25.362

[72] Molina-Sánchez A, Wirtz L. Phonons in single-layer and few-layer MoS_2 and WS_2. Physical Review B: Condensed Matter and Materials Physics. 2011;**84**:1-8. DOI: 10.1103/PhysRevB.84.155413

[73] Zhao W, Ghorannevis Z, Amara KK, Pang JR, Toh M, Zhang X, et al. Lattice dynamics in mono- and few-layer sheets of WS_2 and WSe_2. Nanoscale. 2013;**5**: 9677. DOI: 10.1039/c3nr03052k

[74] Berkdemir A, Gutiérrez HR, Botello-Méndez AR, Perea-López N, Elías AL, Chia C-I, et al. Identification of individual and few layers of WS_2 using Raman spectroscopy. Scientific Reports. 2013;**3**:1755. DOI: 10.1038/srep01755

[75] Pathak CS, Singh JP, Singh R. Effect of dimethyl sulfoxide on the electrical properties of PEDOT:PSS/n-Si heterojunction diodes. Current Applied Physics. 2015;**15**:528-534. DOI: 10.1016/j.cap.2015.01.020

[76] Pathak CS, Singh JP, Singh R. Modification of electrical properties of PEDOT:PSS/p-Si heterojunction diodes by doping with dimethyl sulfoxide. Chemical Physics Letters. 2016;**652**: 162-166. DOI: 10.1016/j.cplett.2016.04.029

[77] Palumbiny CM, Heller C, Schaffer CJ, Körstgens V, Santoro G, Roth SV, et al. Molecular reorientation and structural changes in cosolvent-treated highly conductive PEDOT:PSS electrodes for flexible indium tin oxide-free organic electronics. Journal of Physical Chemistry C. 2014; **118**:13598-13606. DOI: 10.1021/jp501540y

[78] Sun Y, Xia Y. Shape-controlled synthesis of gold and silver nanoparticles. Science. 2002;**298**(5601): 2176. DOI: 10.1126/science.1077229

[79] Sun K, Zhang S, Li P, Xia Y, Zhang X, Du D, et al. Review on application of PEDOTs and PEDOT:PSS in energy conversion and storage devices. Journal of Materials Science: Materials in Electronics. 2015;**26**: 4438-4462. DOI: 10.1007/s10854-015-2895-5

[80] He L, Jiang C, Wang H, Lai D, Rusli. High efficiency planar Si/organic heterojunction hybrid solar cells. Applied Physics Letters. 2012;**100**: 073503. DOI: 10.1063/1.3684872

[81] Xia Y, Zhang H, Ouyang J. Highly conductive PEDOT:PSS films prepared through a treatment with zwitterions and their application in polymer photovoltaic cells. Journal of Materials Chemistry. 2010;**20**:9740-9747. DOI: 10.1039/c0jm01593h

[82] Chiu WW, Travaš-Sejdić J, Cooney RP, Bowmaker GA. Studies of dopant effects in poly(3,4-ethylenedioxythiophene) using Raman spectroscopy. Journal of Raman Specroscopy. 2006;**37**:1354-1361. DOI: 10.1002/jrs.1545

[83] Tran-Van F, Garreau S, Louarn G, Froyer G, Chevrot C. Fully undoped and soluble oligo(3,4-ethylenedioxythiophene)s: Spectroscopic study and electrochemical characterization. Journal of Materials Chemistry. 2001;**11**:1378-1382. DOI: 10.1039/b100033k

[84] Garreau S, Louarn G, Buisson JP, Froyer G, Lefrant S. In situ spectroelectrochemical Raman studies of poly(3,4-ethylenedioxythiophene) (PEDT). Macromolecules. 1999;**32**: 6807-6812. DOI: 10.1021/ma9905674

[85] Sharma B, Fernanda Cardinal M, Kleinman SL, Greeneltch NG, Frontiera RR, Blaber MG, et al. High-performance SERS substrates: Advances and challenges. MRS Bulletin. 2013;**38**: 615-624. DOI: 10.1557/mrs.2013.161

[86] Ko H, Singamaneni S, Tsukruk VV. Nanostructured surfaces and assemblies as SERS media. Small. 2008;**4**:1576-1599. DOI: 10.1002/smll.200800337

[87] Lombardi JR, Birke RL, Lu T, Xu J. Charge-transfer theory of surface enhanced Raman spectroscopy. The Journal of Chemical Physics. 1986;**84**: 4174-4180. DOI: 10.1063/1.450037

[88] Li X, Zhang J, Xu W, Jia H, Wang X, Yang B, et al. Mercaptoacetic acid-capped silver nanoparticles colloid: Formation, morphology, and SERS

activity. Langmuir. 2003;**19**:4285-4290. DOI: 10.1021/la0341815

[89] Hao Z, Mansuer M, Guo Y, Zhu Z, Wang X. Ag-nanoparticles on UF-microsphere as an ultrasensitive SERS substrate with unique features for rhodamine 6G detection. Talanta. 2016; **146**:533-539. DOI: 10.1016/j.talanta.2015.09.024

[90] Bosnick K, Maillard M, Brus L. Single molecule Raman spectroscopy at the junctions of large Ag nanocrystals. The Journal of Physical Chemistry. B. 2003;**107**:9964-9972. DOI: 10.1021/jp034632u

[91] Tao A, Sinsermsuksakul P, Yang P. Tunable plasmonic lattices of silver nanocrystals. Nature Nanotechnology. 2007;**2**:435-440. DOI: 10.1038/nnano.2007.189

[92] Mulvihill MJ, Ling XY, Henzie J, Yang P. Anisotropic etching of silver nanoparticles for plasmonic structures capable of single-particle SERS. Journal of the American Chemical Society. 2010;**132**:268-274. DOI: 10.1021/ja906954f

[93] Jain PK, Lee KS, El-Sayed IH, El-Sayed MA. Calculated absorption and scattering properties of gold nanoparticles of different size, shape, and composition: Applications in biological imaging and biomedicine. The Journal of Physical Chemistry. B. 2006; **110**:7238-7248. DOI: 10.1021/jp057170o

[94] Mahmoud MA, El-Sayed MA. Different plasmon sensing behavior of silver and gold nanorods. Journal of Physical Chemistry Letters. 2013;**4**: 1541-1545. DOI: 10.1021/jz4005015

[95] Huang X, Neretina S, El-Sayed MA. Gold nanorods: From synthesis and properties to biological and biomedical applications. Advanced Materials. 2009; **21**:4880-4910. DOI: 10.1002/adma.200802789

[96] Halas NJ, Moskovits M. Surface-enhanced Raman spectroscopy: Substrates and materials for research and applications. MRS Bulletin. 2013;**38**: 607-611. DOI: 10.1557/mrs.2013.156

[97] Le F, Brandl DW, Urzhumov YA, Wang H, Kundu J, Halas NJ, et al. Metallic nanoparticle arrays: A common substrate for both surface-enhanced Raman scattering and surface-enhanced infrared absorption. ACS Nano. 2008;**2**: 707-718. DOI: 10.1021/nn800047e

[98] Fan JA, Wu C, Bao K, Bao J, Bardhan R, Halas NJ, et al. Self-assembled plasmonic nanoparticle clusters. Science. 2010;**328**:1135-1138. DOI: 10.1126/science.1187949

[99] Kho KW, Shen ZX, Zeng HC, Soo KC, Olivo M. Deposition method for preparing SERS-active gold nanoparticle substrates. Analytical Chemistry. 2005;**77**:7462-7471. DOI: 10.1021/ac050437v

[100] Lee CH, Hankus ME, Tian L, Pellegrino PM, Singamaneni S. Highly sensitive surface enhanced Raman scattering substrates based on filter paper loaded with plasmonic nanostructures. Analytical Chemistry. 2011;**83**:8953-8958. DOI: 10.1021/ac2016882

[101] Qu L-L, Li D-W, Xue J-Q, Zhai W-L, Fossey JS, Long Y-T, et al. Batch fabrication of disposable screen printed SERS arrays. Lab on a Chip. 2012;**12**: 876-881. DOI: 10.1039/C2LC20926H

[102] Yu WW, White IM. A simple filter-based approach to surface enhanced Raman spectroscopy for trace chemical detection. The Analyst. 2012; **137**:1168. DOI: 10.1039/c2an15947c

[103] Yu WW, White IM. Inkjet printed surface enhanced Raman spectroscopy array on cellulose paper. Analytical Chemistry. 2010;**82**:9626-9630. DOI: 10.1021/ac102475k

[104] Hulteen JC, Van Duyne RP. Nanosphere lithography: A materials general fabrication process for periodic particle array surfaces. Journal of Vacuum Science and Technology A. 1995;**13**:1553

[105] Zhang X, Whitney AV, Zhao J, Hicks EM, Van Duyne RP. Advances in contemporary nanosphere lithographic techniques. Journal of Nanoscience and Nanotechnology. 2006;**6**:1920-1934. DOI: 10.1166/jnn.2006.322

[106] Singh DP, Kumar S, Singh JP. Morphology dependent surface enhanced fluorescence study on silver nanorod arrays fabricated by glancing angle deposition. RSC Advances. 2015;**5**: 31341-31346. DOI: 10.1039/c5ra03225c

[107] Dick LA, McFarland AD, Haynes CL, Van Duyne RP. Metal film over nanosphere (MFON) electrodes for surface-enhanced Raman spectroscopy (SERS): Improvements in surface nanostructure stability and suppression of irreversible loss. The Journal of Physical Chemistry. B. 2002;**106**: 853-860. DOI: 10.1021/jp013638l

[108] Cinel NA, Cakmakyapan S, Butun S, Ertas G, Ozbay E. E-beam lithography designed substrates for surface enhanced Raman spectroscopy. Photonics and Nanostructures - Fundamentals and Applications. 2015; **15**:109-115. DOI: 10.1016/j.photonics.2014.11.003

[109] Jeong H-H, Mark AG, Gibbs JG, Reindl T, Waizmann U, Weis J, et al. Shape control in wafer-based aperiodic 3D nanostructures. Nanotechnology. 2014;**25**:235302. DOI: 10.1088/0957-4484/25/23/235302

[110] Gunnarsson L, Bjerneld EJ, Xu H, Petronis S, Kasemo B, Käll M. Interparticle coupling effects in nanofabricated substrates for surface-enhanced Raman scattering. Applied Physics Letters. 2001;**78**:802-804. DOI: 10.1063/1.1344225

[111] Fan M, Andrade GFS, Brolo AG. A review on the fabrication of substrates for surface enhanced Raman spectroscopy and their applications in analytical chemistry. Analytica Chimica Acta. 2011;**693**:7-25. DOI: 10.1016/j.aca.2011.03.002

[112] Le Ru EC, Etchegoin PG, Grand J, Félidj N, Aubard J, Lévi G, et al. Surface enhanced Raman spectroscopy on nanolithography-prepared substrates. Current Applied Physics. 2008;**8**: 467-470. DOI: 10.1016/j.cap.2007.10.073

[113] Kahl M, Voges E, Kostrewa S, Viets C, Hill W. Periodically structured metallic substrates for SERS. Sensors and Actuators B: Chemical. 1998;**51**: 285-291. DOI: 10.1016/S0925-4005(98)00219-6

[114] Hatab NA, Hsueh C-H, Gaddis AL, Retterer ST, Li J-H, Eres G, et al. Free-standing optical gold bowtie nanoantenna with variable gap size for enhanced Raman spectroscopy. Nano Letters. 2010;**10**:4952-4955. DOI: 10.1021/nl102963g

[115] Jen Y-J, Huang J-W, Liu W-C, Chan S, Tseng C-H. Glancing angle deposited gold nanohelix arrays on smooth glass as three-dimensional SERS substrates. Optical Materials Express. 2016;**6**:697. DOI: 10.1364/OME.6.000697

[116] Jen Y-J, Chan S, Huang J-W, Jheng C-Y, Liu W-C. Self-shadowing deposited pure metal nanohelix arrays and SERS application. Nanoscale Research Letters. 2015;**10**:498. DOI: 10.1186/s11671-015-1205-8

[117] Han Y-A, Ju J, Yoon Y, Kim S-M. Fabrication of cost-effective surface enhanced Raman spectroscopy substrate using glancing angle deposition for the detection of urea in body fluid. Journal of Nanoscience and Nanotechnology. 2014;**14**:3797-3799. DOI: 10.1166/jnn.2014.8184

[118] Abdulhalim I. Plasmonic sensing using metallic nano-sculptured thin films. Small. 2014;**10**:3499-3514. DOI: 10.1002/smll.201303181

[119] Singh JP, Chu H, Abell J, Tripp RA, Zhao Y. Flexible and mechanical strain resistant large area SERS active substrates. Nanoscale. 2012;**4**: 3410-3414. DOI: 10.1039/c2nr00020b

[120] Rajput A, Kumar S, Singh JP. Vertically standing nanoporous Al-Ag zig-zag silver nanorod arrays for highly active SERS substrates. Analyst. 2017; **142**:3959-3966. DOI: 10.1039/C7AN00851A

[121] Namura K, Imafuku S, Kumar S, Nakajima K, Sakakura M, Suzuki M. Direction control of quasi-stokeslet induced by thermoplasmonic heating of a water vapor microbubble. Scientific Reports. 2019;**9**:4770. DOI: 10.1038/s41598-019-41255-5

[122] Kumar S, Doi Y, Namura K, Suzuki M. Plasmonic nanoslit arrays fabricated by serial bideposition: Optical and surface-enhanced Raman scattering study. ACS Applied Bio Materials. 2020. Article ASAP, DOI: 10.1021/acsabm.0c00215 [Accessed: 21 April 2020]

[123] Semin DJ, Rowlen KL. Influence of vapor deposition parameters on SERS active Ag film morphology and optical properties. Analytical Chemistry. 1994; **66**:4324-4331. DOI: 10.1021/ac00095a032

[124] Chaney SB, Shanmukh S, Dluhy RA, Zhao YP. Aligned silver nanorod arrays produce high sensitivity surface-enhanced Raman spectroscopy substrates. Applied Physics Letters. 2005;**87**:031908. DOI: 10.1063/1.1988980

[125] Zhou Q, Li Z, Yang Y, Zhang Z. Arrays of aligned, single crystalline silver nanorods for trace amount detection. Journal of Physics D: Applied Physics. 2008;**41**:152007. DOI: 10.1088/0022-3727/41/15/152007

[126] Zhang ZY, Zhao YP. Tuning the optical absorption properties of Ag nanorods by their topologic shapes: A discrete dipole approximation calculation. Applied Physics Letters. 2006;**89**:23-25. DOI: 10.1063/1.2221403

[127] He Y, Larsen G, Li X, Ingram W, Chen F, Zhao Y. Nanoscale conical swiss roll with broadband visible and nir circular dichroism. Advanced Optical Materials. 2015;**3**:342-346. DOI: 10.1002/adom.201400435

[128] Mark AG, Gibbs JG, Lee TC, Fischer P. Hybrid nanocolloids with programmed three-dimensional shape and material composition. Nature Materials. 2013;**12**:802-807. DOI: 10.1038/nmat3685

[129] Titus J, Larsen G, Zhao Y, Perera AGU. Large circular dichroism and optical rotation in titanium doped chiral silver nanorods. Annals of Physics. 2016;**528**:677-683. DOI: 10.1002/andp.201600103

[130] Nair G, Singh HJ, Ghosh A. Tuning the chiro-plasmonic response using high refractive index-dielectric templates. Journal of Materials Chemistry C. 2015; **3**:6831-6835. DOI: 10.1039/C5TC00922G

[131] Nair G, Singh HJ, Paria D, Venkatapathi M, Ghosh A. Plasmonic interactions at close proximity in chiral geometries: Route toward broadband chiroptical response and giant enantiomeric sensitivity. Journal of Physical Chemistry C. 2014;**118**: 4991-4997. DOI: 10.1021/jp4117535